[ECO]systems of Resilience Practices

[ECO]systems of Resilience Practices

Contributions for Sustainability and Climate Change Adaptation

Edited by

ANGELA COLUCCI
REsilienceLAB/Co.O.Pe.Ra.Te. Ldt, Italy

GIULIA PESARO
REsilienceLAB/Co.O.Pe.Ra.Te. Ldt, Italy

ELSEVIER

Elsevier
Radarweg 29, PO Box 211, 1000 AE Amsterdam, Netherlands
The Boulevard, Langford Lane, Kidlington, Oxford OX5 1GB, United Kingdom
50 Hampshire Street, 5th Floor, Cambridge, MA 02139, United States

Notices
Knowledge and best practice in this field are constantly changing. As new research and experience broaden our understanding, changes in research methods, professional practices, or medical treatment may become necessary.

Practitioners and researchers must always rely on their own experience and knowledge in evaluating and using any information, methods, compounds, or experiments described herein. In using such information or methods they should be mindful of their own safety and the safety of others, including parties for whom they have a professional responsibility.

To the fullest extent of the law, neither the Publisher nor the authors, contributors, or editors, assume any liability for any injury and/or damage to persons or property as a matter of products liability, negligence or otherwise, or from any use or operation of any methods, products, instructions, or ideas contained in the material herein.

ISBN: 978-0-12-819198-9

Publisher: Candice Janco
Acquisitions Editor: Marisa Lafleur
Editorial Project Manager: Jai Marie Jose
Production Project Manager: Joy Christel Neumarin Honest Thangiah
Cover Designer: Victoria Pearson

Typeset by MPS Limited, Chennai, India

Contents

Part III Exploring resilience practices

List of contributors

Jerry Andriessen
Wise & Munro, Den Haag, The Netherlands

Nicolas Beaurez
Cerema, France

Federica Benelli
Poliedra–Politecnico di Milano, Italy

Lise Bourdeau-Lepage
CNRS UMR EVS, Université de Lyon, Lyon, France

Elis Tayná Almeida Brazil
Ateliê de Arquitetura e Urbanismo, Cidade Universitária, Universidade Federal do Pará, Belém, Brazil

Ana Cláudia Duarte Cardoso
PGITEC, Cidade Universitária, Universidade Federal do Pará, Belém, Brazil

Lisa Cassanet
Red de Transición, Spain

Luca Cetara
Eurac Research

Angela Colucci
ResilienceLAB/Co.O.Pe.Ra.Te. Ldt, Italy

Mara Cossu
Poliedra–Politecnico di Milano, Italy

Marie-Laure Couderc-Papaix
Cerema, France

C.S. Dechamma
Community Engagement Initiative, Azim Premji University, Bangalore, Karnataka, India

Juan Del Río
Red de Transición, Spain; ECOLISE, Spain

Catherine Dezio
Department of Architecture and Urban Studies, Politecnico di Milano, Milan, Italy

Nicolas Furmanek
Cerema, France

Valeria Garibaldi
Fondazione Cariplo, Milan, Italy

Salvador Garrido-Soler
Red de Transición, Spain

Elena Jachia
Fondazione Cariplo, Milan, Italy

Kateřina Sidiropulu Janků
Czech Technical University in Prague, Czechia; Carinthia University of Applied Sciences, Spittal an der Drau, Austria

Eliot Laniado
Poliedra–Politecnico di Milano, Italy

Sarah Talandier Lespinasse
Cerema, France

José Júlio Ferreira Lima
PGITEC, Cidade Universitária, Universidade Federal do Pará, Belém, Brazil

Stefano Magaudda
U-Space srl, Italy

Marcello Magoni
Climate Change, Risk and Resilience Laboratory, Department of Architecture and Urban Studies (DAStU), Politecnico di Milano, Milan, Italy

Andrea Mangiatordi
University of Milano-Bicocca, Milano, Italy

Davide Marino
Department of Biosciences and Territory, University of Molise, Pesche, Italy

Giampiero Mazzocchi
Department of Policies and Bio-economics, Council for Agricultural Research and the Analysis of Agricultural Economics, Rome, Italy

Thales Barroso Miranda
Secretariat of Environment of the City of Belém, Belém, Brazil

Cristian Moyano-Fernández
Red de Transición, Spain; Universitat Autònoma de Barcelona

Serena Muccitelli
U-Space srl, Italy

Harini Nagendra
Azim Premji University, Bangalore, Karnataka, India

Mirjam Pardijs
Wise & Munro, Den Haag, The Netherlands

Giulia Pesaro
REsilienceLAB/Co.O.Pe.Ra.Te. Ldt, Italy

Rachele Radaelli
Climate Change, Risk and Resilience Laboratory, Department of Architecture and Urban Studies (DAStU), Politecnico di Milano, Milan, Italy

Guillermo Rodriguez-Navas
Red de Transición, Spain

Amrita Sen
Department of Humanities and Social Sciences, Indian Institute of Technology Kharagpur, West Bengal, India; Azim Premji University, Bangalore, Karnataka, India

Carlotta Sigismondi
Poliedra–Politecnico di Milano, Italy

Foreword: Rethinking resilience practices before and after the Covid-19 crisis

In recent years, there has been a remarkable expansion in the use of the resilience concept and a parallel wider diffusion of resilience practices, making territorial and societal quality a central element in urban and territorial innovative actions and investments. Moreover, looking at recent years' experiences, the concept of resilience seems particularly effective in mobilizing local resources and funding around the community and territorial-based projects to enhance the quality of life and the natural environment.

The authors' contributions enrich and follow this overall framework. The sharing of experiences and reflections aims to enhance the contribution of scientific thinking, knowledge building, and reframing for improving territorial and community resilience as both a crucial territorial quality and an overall sustainable development goal.

The book has been conceived before the world was overwhelmed by the crisis caused by Covid 19, and the contributions, of course, do not mention it directly. On the other hand, resilience has been one of the most popular words at the global level during the whole last year. So popular that there has been a severe risk to make it lose its meaning and power.

Look at the experiences and practices described in the book, aiming at enhancing resilience at the local level and preparing people to better cope with crises and hazards. The narrative of communities' adaptation efforts and the resources, knowledge, know-how, tools, and means to support capacity building and the diffusion of adaptation behaviors is often not so much different from that of the stories and experiences shared during the pandemic. Therefore we invite readers to draw attention to some elements and dynamics that could be assumed as inspiring references to understand how territories and communities can respond to the impacts of the crisis and increase their resilience.

To complete the reasoning, we add here some common consideration elements about the impacts of Covid 19 and the lessons learned during such a challenging period.

One of the most cross-cutting and common elements highlighted by the crisis has been the direct and indirect *involvement of communities and individuals in mutual support* networks to better face the pandemic's impacts in everyday life and overcome difficulties in the supply of essential goods and services. The availability of previously existing networks, together with the capability to establish new ones when needed, was revealed to be an essential resource for community resilience. There have been specific local practices activated in many places all over the world revealing how much people need social and mutual networking not only looking at specific solutions to face specific problems but also to make individuals and families feel not alone and part of a community, so far coupling practical and psychological help. Likewise, there has been a recovery of contact with the neighborhood community, there has been greater visibility of a sincere concern for family and loved ones, and in general, the bonds of solidarity toward certain groups (those most affected by the pandemic) have been extended.

The organization itself of societies and groups, on the other hand, is one of the elements explaining the diffusion of the infection. In many places, especially (but not only) in the developing countries, the infection has spread around because of a mix of social dynamics and poor quality and accessibility conditions of the infrastructures and public services organization. For instance, the interactions linked to the provision of services and domestic work provided by low-income social groups can explain some specific infection trends. The lack of infrastructure endowment and quality in low-income urban settlements (in particular of water supply) demonstrated how an essential resource is crucial in preventing hygiene measures and disease diffusion. Moreover, the strong dependence on commercial activities to generate income for the population, especially the work in local fairs and markets, often reflects the intense urban—rural interactions that strongly contributed to the local spread of the pandemic. Similar reflection can be related to the dependence on city-based health care or the everyday supplies to an urban population that has revealed itself as highly vulnerable.

The fast change, particularly during the first wave, clearly demonstrated the fundamental *importance of resilience capabilities at the individual and social levels*. The pandemic has besieged countries worldwide, and all typologies of society and people have found themselves exposed to the same event simultaneously. On the other hand, it has revealed different vulnerability degrees and fragilities in different population groups and individuals. For example, the situation of isolation and the pressure, in some cases, lived as coercion. Adapting to distance relationships and exchanges (when and if possible) for quite all the working and not-working time has generated a considerable health impact and psychological cost. Many people have become exhausted from virtual and teleworking tools, and the need to maintain physical human contact and go outdoors has become evident. This emotional stress has been compounded by the fear caused by the possibility of catching the virus, the suspension of certain freedoms, the economic instability, and the fragmentation of citizen security. Precisely because of these factors, inner transition work to alleviate this fatigue, anxiety, fear, or frustration is becoming increasingly meaningful, while it is now more evident that the inner transition approach is not replaceable by the outer transition. In addition, the pandemic context has shown that one of the groups most affected by the conditions of confinement and social distance has been the children. That phenomenon so impacting this specific population sector has solicited the movements working on resilience enhancement at the community level to embrace more explicitly the youngest, helping them develop remote tools and outer practices and strengthening their capacities for emotional resilience.

The intense stress on our lives has intensified (or made more evident) the tension concerning the psychological—social resilience between the individual resilience, which is more potent/widespread because rooted and consolidated in psychology and the community resilience.

The pandemic event was a period of great stress on the individual dimension and has further strengthened the focus on the "human" or sociopsychological resilience in its "individual" dimension.

A crucial effort, both during the pandemic period and in the recovery phases, has been and must be placed in strengthening the dimension of the community resilience as an important area both for strengthening the resilience capacities of territorial systems and guaranteeing benefits to individuals.

It is worthwhile to highlight also to what extent the availability of means and technologies to overcome some of the limits imposed by the pandemic have become, at the same time, sources of stress for different reasons. The need to quickly adapt to ICT and digital technologies as the only or significant ways to teach and learn, buy and supply, produce and consume, work and spend leisure time, and even contact loved ones created ambivalent reactions. The feeling of being able to overcome the contingent problems has been many times balanced by the fear for the need to abandon the comfort zones of the previous normality and past lifestyles. *Resilience implies the capacity to learning from coping and overcoming the shock or trauma.* In this case, few lessons seem been learned: achieving a new balance, as in the resilience thinking, is still to come, overshadowed by the need for many to return too fast to the past instead of confronting the new challenges post—Covid era.

At the same time, another critical impact has occurred during the pandemic. Online and remote working has become more widespread and improved, normalizing and making many collective activities more effective. Moreover, ICTs have been able to produce extraordinary benefits by looking at the participation of people in meetings, workshops, and other activities offered in distance mode, allowing large audiences to be involved in initiatives and dissemination about various issues, among which resilience, sustainability, and green. It has also been an opportunity for transition initiatives, most of which are group-based, to continue meeting and innovating with remote transition strategies.

This crisis has other specificities. Being people confined because of the risk of infection, nature and the activities related to nature have, to some extent, found relief from the high pressures produced by the everyday life dynamics of contemporary societies. The quality of natural resources has increased quite everywhere globally, and the benefits have been visible. On the other hand, a different relationship of citizens with nature and the role itself of nature as a relief for people. *Nature and the natural environments have therefore been recognized as resilience resources.* Some activities, such as spending time in green areas inside and outside cities and growing one's food, have become common among people who have experienced confinement within their homes, whether large houses with gardens or small flats. Beyond having a good time and keeping busy doing different things, activities such as these may have evidenced a psychological or spiritual need to maintain contact with nature and get positive effects from observing vegetables growing. These practices launched and explored during this period can be premises for more space and recognition, in the future, for the development of projects and activities based on nature protection and valorization, and on the quality of natural resources as tools for territorial and community resilience enhancement. Finally, these dynamics can also produce positive externalities on some economic activities, like food production. Agroecological food producers started to deliver food at home or enhanced already existing delivering services in many countries, contributing to the permanence of people at home and expanding the network of producers and their possibilities of generating income. At least in Europe, there is also evidence that organic food consumption has increased, thanks to an improvement of online commerce, thus increasing the farmers' revenues and making the activities linked to organic and zero kilometers agriculture more attractive.

The mutual benefit for the well-being of nature and human

During the confinement, all of us have experienced how the presence of a piece of green or nature in our artificialized systems (not only urban) has been crucial for our well-being. From the media and our experiences, the necessity for a direct relationship with nature was evident.

The pandemic (leaving in the background the reflections on the causes linked to the origin of the pandemic itself) has made urgent a reflection around well-being and mutual relations of stewardship between human/community/nature.

Well-being concerning the individual in drawing direct benefit: frequenting/staying in the green and nature and taking care of the green and therefore an intimate relationship of mutual benefit in the care that produces well-being. Many of the practices of care and action in the green mentioned above are based on this intimate relationship.

When a community takes care of green spaces, strengthening at the same time the social cohesion the Well-being, and mutual benefits are exponentially incremented. The multiple benefits from this relationship are the individual well-being and collective benefits in strengthening social cohesion and improving community places (identity with a place). Collective well-being also refers to ecosystem conservation that can provide ecosystem services (purification of air).

The book focuses on the collective dimension of the mutual benefits deriving from the mutual stewardship between community and nature (ecosystems and green care practices), intending to underline the importance of community action in reconstructing these fundamental benefits to systems territorial and also aware of the fundamental and intimate individual relationship.

This unprecedented crisis also revealed some *vulnerabilities of contemporary economies*. Covid 19 can be regarded as an international level disaster but, differently from other natural catastrophes, it did not destroy the production and consumption resources themselves, the buildings hosting commercial activities, or the infrastructures. It reduces the human exchanges tremendously, thus making commercial and other services sectors particularly exposed and vulnerable. Moreover, the high complexity of contemporary societies and the variety of activities, and the related dynamics become part of the problems that have been particularly evident in difficulty to take into consideration the whole of the economic activities to be considered for the design and implementation of economic support intervention by the national governments. The complexity of relationships, sharing dynamics, and exchanges need to be better understood to increase the resilience capabilities of the different categories of territorial subjects according to their roles and functions. This in the light of the evidence that the Covid crisis has lead to stable changes, like those in working models and online activities, made possible by a quite rapid enhancement and evolution of ICT and digital services. It requires profoundly researching territorial and community sustainability and reasoning about a possible rebalancing of regions and the urban–rural and local–over local interactions.

There are hundreds of thousands of possible examples of how collectivities and individuals have organized themselves worldwide during the Covid-19 crisis. As in "normal times," what is essential for future development is to learn from the practices and stabilize the new solutions and the created capabilities and innovation as a heritage for the future. The legacy in terms of resources, activities, creativity, and improvements comes from the crisis, allowing societies to move toward new and renewed ways of making the territorial system and communities more sustainable and resilient. A legacy of experiences to be treasured from all viewpoints, from the functioning of our production and consumption models to the impacts they have on our natural environments and their equilibriums up to the new/renewed centrality of the values attached to the availability and quality of all kinds of public services and proactive behaviors and involvement of collectivities and organizations at all territorial scales, starting from the local. A heritage of practices to be carefully studied and understood so to highlight the most important values and resources a community should preserve and continuously regenerate—material and immaterial ones, public and private ones, related to nature and culture protection, to welfare and health, to new networking and connections building—as a precious territorial capital.

We hope that these reflections could address the reading of the book toward an enhanced understanding of what resilience is and to what extent making the concept of territorial and community resilience operational could produce not only new ideas and resources for better facing future crisis but also more diffused good practices and behaviors to enhance well-being and sustainability.

Giulia Pesaro, Angela Colucci, Elis Tayná Almeida Brazil, Thales Barroso Miranda, C.S. Dechamma, Juan del Río, Ana Cláudia Duarte Cardoso, José Júlio Ferreira Lima, Salvador Garrido Soler, Cristian Moyano Fernández, Harini Nagendra, Guillermo Rodriguez-Navas and Amrita Sen

Introduction: Resilience, climate change, and sustainability in practice from approaches to action

Angela Colucci and Giulia Pesaro
REsilienceLAB/Co.O.Pe.Ra.Te. Ldt, Italy

1 Resilience in action

In the last decade, the concept and the term of resilience have had an enormous diffusion. Environmental, social, and economic urgencies have prompted the spread of a powerful concept such as resilience, which allows a positive vision in recovery, adaptation, and evolution dealing with global and local stress and shocks (Berkes et al., 2000; Folke & Gunderson, 2010). The diffusive use of the concept as a metaphor to activate territorial and urban strategies has undoubtedly led to an advancement of the debate on the theoretical framework and tools for action (Galderisi & Colucci, 2018).

The resilience concept applied to ecological systems behavior appeared in the literature debate in 1973 when Holling published an article on the persistence of ecosystems in absorbing disturbance maintaining recognizability and functionality factors (Holling, 1973). Resilience assumes different meanings and conceptualizations in relation to the disciplinary fields frames (Colucci, 2012) among which it is possible to recognize shared capacities of the systems to cope with stress and shocks. Social science developed the psychological definition of resilience, focusing on the capacities of communities to cope with disturbances and adaptation to changes at local–global levels.

In literature, several authors enlighten the overuse of the resilience concept that often implies an overlay and confusion in proposes producing an ambiguity and decreasing the value and complexity of resilience concept itself (Brand & Jax, 2007).

The book aims to embrace different resilience conceptualizations deriving from different disciplinary approaches focusing on the resilience capacities enchantment, promoted through processes and actions launched and managed by multistakeholders' partnerships.

Moreover, the focus is on resilience practices by communities and territories at the local level. The analysis processes aim therefore to better understand how and in which ways resilience become an operational concept, a concept interpreted, developed, customized, and adapted in relation to the local specificities, according to the peculiar demand for action and the construction of skills and capabilities to face present and future crises.

This is why, according to the specific focuses addressed by the authors, the contributions recall the definition of the concept of resilience which is better suited to the specificities of the case treated. A variety and diversity will be *solved* in the closing chapters, where a reframing of the concept will be proposed with alignments and new perspectives for action as operational inputs in the literature debate on resilience applied to the territorial contexts (Burayidi et al., 2020; Davoudi et al., 2013).

The focus of this book is on the resilience practices intended as processes driven (and acted) by multiple actors (communities and civil society actors) and in which activities aim to improve (innovate) services, approaches, routines, practices, or infrastructures existing in local and territorial contexts. This broad definition of resilience practices includes the transition initiatives, grassroots and an extensive range of local community-led initiatives that could be clustered in differentiated and flexible geographies concerning the governance of processes, the main focus (issue or lever for mobilization activation), and the action tools (Colucci, 2018; Frantzeskaki et al., 2016). Practices are in general acting on "commons": interventions activated are oriented to the improvement of public spaces, natural and built public environments, common goods, and services through community-led actions generating positive benefits on environmental pressure phenomena (related to climate change) and on community strengthening. The central orientation to common values, public interests, and (global) sustainable goals can be identified as a file-rouge and core criteria in selecting the practices. The Ostrom (Ostrom, 2009) approach and contributions on the self-organized system, having frequently solved many commons dilemmas through action and creativity, support the crosscutting reframe concepts navigating the book chapters and boosting the emergence of synergic forces and dynamics which revealed to be important conditions for the success of the initiatives put in place all over the world.

Along the chapters and cases presented, three main nodes of tensions emerge connecting the resilience practices exploration to the literature debate: time dimensions in resilience re-action; intentional design of "resilience"; and mainstreaming, diffusion, and complexity reduction.

Time dimensions in resilience re-action. Recovery, adaptive, and evolutionary resilience imply different time perspectives and different assumptions of resilience concept definition and often also conflictual strategies and solutions. As several authors suggested (Davoudi et al., 2012; Davoudi et al., 2013) the three dimensions of resilience unavoidably coexist in the territorial/urban transformation processes (project/planning, and interventions), providing richness and complexity with the condition these have and have to be consciously managed and directed. Resilience practices do not always demonstrate awareness of potential dyscrasia points along the resilience temporalities and the reference approaches but often reveled to be able to implicitly manage the adaptive and evolutionary resilience.

Intentional design of resilience. Another tension that is widely discussed in the literature but often not considered in "practices" processes (and in media debate) is if it is possible to "plan resilience or design resilience" or if it is more coherent with the resilience concept (capacities of a complex system) to design/plan strategies and solution for the enhancement of resilience capacities of a complex system. In resilience practices action, the tension due to "imaginative" and "design" thinking is often implicit, as well as the rational assumption of other core nodes often animating the literature debate about the resilience dilemma (such as resilience of who/what and for who/to whom) (Davoudi et al., 2013).

Mainstreaming, diffusion, and complexity reduction. The diffusion of resilience implied an enormous production and diffusion of "receipts" and solution toolkits often reducing the complexity and making the resilience "easier and simpler." In the disciplinary debate in the territorial planning and design fields the tension along the needs to provide "toolkits" for implementation of sustainable transition and the risk of missing crucial values and principles in reducing the intrinsic complexity of the approach is not new. The mainstreaming and diffusion of collective and individual practices, actions, and behaviors able to cope with climate, resilience, and sustainability challenges have to be balanced with the necessity to maintain a long-term visioning rooted in complex thinking and approach. The challenge is therefore how to foster the complex and powerful resilience concept in feasible and implementable practices and interventions without the risk of a substantial simplification and impoverishing of action potentialities.

1.1 Resilience in action: [eco]systems of practices

The book and all contributions refer to resilience specifically addressing resilience in action and act, which allows to highlight the potentialities and responsibilities of people in their territories and to better, explore and discuss the role and contribution of the grassroots or community-led process. The increased attention of literature on climate change, social innovation, social cohesion, the launch of diversified worldwide networking platforms, and the media's attention demonstrate a worldwide process recognizing the contribution of grassroots initiatives and the improved role of local communities and citizens' commitment. The launch of web platforms collecting seed and good practices has become very common, together with the increase in relevance and quality of transition initiatives networking (e.g., the Transition movement), providing visibility but also mutual tools and support for local communities and citizens group. These dynamics imply the need for a profound reflection on the role of the practices and their contribution in the achievement of long-term sustainable goals and in the enhancement of territorial systems resilience capacities, with specific reference to the legitimacy (ethical and democracy) of these interventions.

This is why, in the book, the different contributions that communities and citizens can provide in improving local and global resilience are specifically addressed and deeply explored. As mentioned above, the authors assume different definitions and approaches to resilience with the aim to highlight both specificities and commonalities of resilience practices focusing on the transformative processes activated by multistakeholders partnerships engaging local communities in the decision-making, implementation, and management of initiatives providing benefits on multiple components of complex systems.

Climate, ecological, social, economic stress phenomena are drivers in the activation of resilience practices that, in facing the specific urgencies, integrate multiple benefits at the local level and contribute in improving the overall quality of living places and conditions.

The acted practices are explored without the strict necessity of organizing them into categories and the main attention is paid to the whole understanding of the barriers which can appear and affect the upscaling and stabilization of practices. Without standardization and homogenization purposes, the aim is to recognize the perspectives and conditions for a shift from a fragmented and heterogenous panorama to a more organic "ecosystem of practices" able to activate synergies and relationships among the different resources, components, energies, and richness which are part of it and enable it to adapt and develop overtime on the basis of its own endogenous strengths.

Approaching the arena of resilience practices, an immense treasure of activities, local knowledge, and operational capabilities emerges which needs to be better understood, stabilized, and strengthened to enhance territorial and community resilience as a renewed intrinsic characteristic of the ecosystem.

Therefore it is crucial to work on the quality of the project design, the availability of specific tools and solutions tailored to local communities' needs, the capacity building related to good governance, and the development of the decision-making and implementation processes. The issues raised are strategic and central to stabilizing and fertilizing return practices, a work path through which it will be possible to enhance the effectiveness and efficiency of the communities' action toward territorial resilience at both the local and system levels. This in the light of the choice for conceiving the contribution to sustainability and climate change mitigation and adaptation is essential for a territorial and community resilience development strategy. In the light of the variety of territorial components, sectors, and actors involved, it is finally urgent to strengthen the contribution of tools and methods based on a multidisciplinary and multisectorial approach and a more stable involvement of all actors.

2 [Eco]system of resilience practices

The first part on *Resilience Practices: Challenges and Emerging Issues* reframes three emerging debate issues on the resilience concept applied to complex systems and previously underlined as one of the tensions characterizing the resilience practices acting: the collaborative and social inclusion aspects, which will be discussed in Chapter 1, and the economic dimension and multiple benefits in approaching long-term actions for resilience capacities improvement, discussed in Chapters 2 and 3. The economic sphere is assumed and recognized as a crucial dimension for the stabilization of practices, which is a precondition for practices contribution to long-term sustainable development goals.

The second part *Resilience Practices Observatory Project* presents the Italian Resilience Practices Observatory (RPO) project introduced by the mission and vision of Cariplo Foundation in supporting the RPO and the Resilient Communities projects. The activities and results of the RPO, (launched in 2015) and the principal phenomena emerging from the resilience practices interpretation and characterization will be introduced in Chapter 4 and deeply analyzed in Chapter 5. The RPO resilience practice database analysis permits to identify, according to the data collection, the main topics, sectors in which the practices concentrate their activities and efforts, the geographies, the contexts of action, and the most relevant and diffused tools for implementing and managing the transformative processes launched. Two chapters present the developed crosscutting investigation paths. In Chapter 6 an evaluation and monitoring framework is proposed and applied to selected resilience practices. In Chapter 7 the synergies among resilience practices and the food policies are investigated, enlightening the different dimensions of territorial resilience through the polyhedral lens of food. A closing chapter, Chapter 8, summarizes outcomes and inputs that have emerged from the RPO project and underlines the potentialities and criticalities for resilience practices stabilization, upscaling, and blooming.

The third part *Exploring Resilience Practices* proposes a selection of national and international practices presenting different approaches to resilience practices and focusing on different core aspects. Chapters 9 and 10 investigate two crucial nodes in fostering urban resilience interventions and in improving urban/metropolitan landscape quality and systemic resilience in Italy and France. The Agro Pontino case study (Italy) focuses on the governance and collaborative process in innovating food and agroecology in periurban areas as lever in launching local resilience policies. The Lyon case study (France) proposes a broad governance framework and implementation instruments to manage multiscale/multilevel policies of urban transformations.

Lessons and crucial inputs emerge from the presentation of the Transition Network and Movement in Spain. Chapter 11 reports the current state of the Transition Movement Initiatives in the country critically retracing the challenges faced in the 13 years of history of the Movement and emphasizes perspectives in fostering the Movement contribution and positive impacts in times of social and ecological emergency.

The role of local communities' and citizens' engagement and commitment in preserving natural resources and local economies under pressure is investigated presenting multifaced cases in the Indian and Brazilian contexts. Chapter 12 focuses on the urban ecological commons' conservation.

In presenting the lake restoration efforts in the city of Bengaluru (India), it underlines the crucial role played by community groups in rejuvenating urban lakes as commons. Chapter 13 investigates existing connections between urban sprawl's morphological patterns and environmental and social vulnerability in Brazil.

The fourth part on *Perspectives in Boosting Practices' Contribution Toward Sustainability, Adaptation, and Resilience* outlines perspectives for resilience practices stabilization and suggests promising conceptual reframing in connecting resilience concept and the richness emerging from the resilience practices contributions explored. Chapter 14 focuses on the economic sphere as a crucial asset in fostering the stabilization of resilience practices. The creative diversity propriety is the Fil Rouge used in emphasizing the explored and future perspective contributions of the multifaced and rich ecosystem of resilience practices, as defined above, for local and global sustainability, adaptation, and resilience challenges. Creative diversity will be discussed in Chapter 15, where the discussion on the different facets of creative diversity demonstrates the relevance of interventions able to enhance the creative diversity of organizational components, which is critically approached in the chapter dedicated to governance. The governance of transformative processes is investigated in Chapter 16, reframing the theoretical dilemma emerging from the literature debate and underlining both the opportunities identified in the resilience practices cases presented in the book and the perspectives to boost the resilience practices networking, ecosystem, and collaborative and adaptive governance models frameworks. The closing chapter, Chapter 17, aims to highlight the advancement in the management and governance of the commons and in collective action (Ostrom, 1990, 2009) and retraces the emerging key domains to boost the resilience practices contribution toward sustainability. The "ecosystem of resilience practices perspective" reaffirms the relevance of strengthening the awareness of acting as an ecosystem of practices by outlining the assets and perspectives necessary to ensure an effective, transformative process to face and overcome climate change and sustainability challenges and resilient territorial and urban systems.

References

Berkes, F., Folke, C., & Colding, J. (2000). *Linking social and ecological systems: Management practices and social mechanisms for building resilience*. Cambridge University Press.

Brand, F. S., & Jax, K. (2007). Focusing the meaning(s) of resilience: resilience as a descriptive concept and a boundary object. *Ecology and Society*, *12*(1). Available from https://www.jstor.org/stable/26267855.

Burayidi, M. A., Allen, A., Twigg, J., & Wamsler, C. (2020). *The Routledge handbook of urban resilience*. Taylor & Francis.

Colucci, A. (2012). Towards resilient cities. Comparing approaches/strategies. *TeMA. Journal of Land Use, Mobility and Environment*, *5*(2), 101–116, Paginazione. Available from https://doi.org/10.6092/1970-9870/921.

Colucci, A. (2018). *The transition approach in urban innovations. Smart, Resilient and Transition Cities* (pp. 19–28). Elsevier. Available from https://doi.org/10.1016/B978-0-12-811477-3.00003-1.

Davoudi, S., Brooks, E., & Mehmood, A. (2013). Evolutionary resilience and strategies for climate adaptation. *Planning Practice and Research*, *28*(3), 307–322. Available from https://doi.org/10.1080/02697459.2013.787695.

Davoudi, S., Shaw, K., Haider, L. J., Quinlan, A. E., Peterson, G. D., Wilkinson, C., Fünfgeld, H., McEvoy, D., Porter, L., & Davoudi, S. (2012). Resilience: A bridging concept or a dead end? "Reframing" resilience: Challenges for planning theory and practice interacting traps: resilience assessment of a pasture management system in Northern Afghanistan urban resilience: what does it mean in planning practice? Resilience as a useful concept for climate change adaptation? The politics of resilience for planning: A cautionary note: Edited by Simin Davoudi and Libby Porter. *Planning Theory & Practice*, *13*(2), 299–333. Available from https://doi.org/10.1080/14649357.2012.677124.

Folke, C., & Gunderson, L. (2010). Resilience and global sustainability. *Ecology and Society*, *15*(4). Available from https://doi.org/10.5751/ES-03954-150443, art43.

Frantzeskaki, N., Kabisch, N., & McPhearson, T. (2016). Advancing urban environmental governance: Understanding theories, practices and processes shaping urban sustainability and resilience. *Environmental Science & Policy*, *62*, 1–6. Available from https://doi.org/10.1016/j.envsci.2016.05.008.

Galderisi, A., & Colucci, A. (2018). *Smart, resilient and transition cities: Emerging approaches and tools for a climate-sensitive urban development*. Elsevier.

Holling, C. S. (1973). Resilience and stability of ecological systems. *Annual Review of Ecology and Systematics*, *4*(1), 1–23. Available from https://doi.org/10.1146/annurev.es.04.110173.000245.

Ostrom, E. (1990). *Governing the commons: The evolution of institutions for collective action* (1st ed.). Cambridge University Press. Available from https://doi.org/10.1017/CBO9780511807763.

Ostrom, E. (2009). *Understanding institutional diversity*. Princeton University Press.

PART I

Resilience practices: challenges and emerging issues

CHAPTER 1

Collaborative agency and sensitive design for social resilience

Kateřina Sidiropulu Janků[1,2], Mirjam Pardijs[3], Andrea Mangiatordi[4] and Jerry Andriessen[3]
[1]Czech Technical University in Prague, Czechia
[2]Carinthia University of Applied Sciences, Spittal an der Drau, Austria
[3]Wise & Munro, Den Haag, The Netherlands
[4]University of Milano-Bicocca, Milano, Italy

1.1 Social resilience: innovating social frameworks in the neoliberal era

Contemporary societies are highly globalized, dense, and therefore intertwined with each other and fragile. Unexpected, fast-spreading developments are now a shared experience of all, no matter whether living in metropolitan or rural landscapes. The digital global network of meanings, accelerated by the growth of third-generation Internet and social media, is influencing local events and the emotionality of daily life in an unprecedented way among populations. Therefore new forms of governing the Anthropocene are popping up, contesting and alternating existing governance frames, from nation states to the highly globalized neoliberal social order or heterarchical forms of social organization. In parallel, what was previously seen as an underdeveloped state of civilization is now more often seen as an alternative way of doing things and a possible inspiration for the future. In this chapter, we will focus on forms of social development and interactions which seem promising in regard to reinforcing social resilience—collaborative agency. Taking into account the massive development of the digital Anthropocene, we will also focus on sensitive design principles that may support collaborative agency in both offline and online layers of social existence. One of the formative texts introducing the term resilience to social science was the collective monography edited by political economist Peter A. Hall and sociologist Michèle Lamont, *Social Resilience in the Neoliberal Era* (Hall & Lamont, 2013). The

[ECO]systems of Resilience Practices
DOI: https://doi.org/10.1016/B978-0-12-819198-9.00003-X

appearance of this comprehensive piece was preceded by numerous studies on resilience in the field of ecology, developmental psychology, and the analysis of disaster (Hall & Lamont 2013, 12). The authors of the monography also acknowledged the importance of neoliberal economic and structural development, which had intensified since the 1980s and resulted in social conditions that call for innovative forms of tackling adversity as well as more resilience. They also kept in mind the high diversification of neoliberal development across the globe and its devastating influence beyond just human societies. Through the rise of new challenges, new opportunities for innovations opened as well, and this led them to introduce the term social resilience to capture new dimensions in resilience research.

What makes social resilience distinct from previous resilience studies is (1) the emphasis on the group (neither individual nor structural) as the coping agent; (2) the more sociological definition of well-being, which is bound to group belonging in a broader sense (Hall & Lamont, 2013, 12); and (3) the more open and relational definition of the longed-for state of existence being reestablished according to Giddens's meaning of social agency (Giddens, 1984) due to resilience competencies:

> *We see resilience in dynamic terms, not as the capacity to return to a prior state but as the achievement of well-being even when that entails significant modifications to behaviour or to the social frameworks that structure and give meaning to behaviour.*
>
> ***Hall and Lamont (2013, 12)***

We would like to put emphasis on the importance of the creative force of resilient competencies. Being resilient means belonging to a social group that is, as a whole, trying to deal with adversity and reestablish well-being. But it also means more than successfully contending with adversity, it means having the capacity to collaboratively create new frameworks and meanings which will be functional in new and sometimes unforeseen social conditions. As social resilience pertains to the social dimension, that is, the intersubjective, new meanings and solutions are cocreated at the group level, and this would not be possible without the skills of collaboration and collaborative agency. When one talks about change, they refer to their environment, the more local as well as the more global practices in which they operate on a daily basis (Dreier, 1999). The individual ability to deal with changes in practices influenced by technology, geopolitics, populism, climate change, war, manipulative media, and many others is clearly limited, and there are good reasons to doubt the resilience of humans to deal with the immense issues at stake. Indeed, a main concern in today's era is how to slow down or reduce the various negative trajectories, many of which are being perpetrated by people and institutions who have the presumed responsibility of working on behalf of collectives. Such tendencies can only be tackled with collective and collaborative action that supports the social resilience of human stakeholders.

This chapter will, through the use of diverse analytical descriptions of real-life situations anchored in European societies, further elaborate social resilience from several conditional elements crucial in facing unforeseen challenges and producing functional solutions. After we have elaborated our view on collaboration for resilience from a theoretical perspective, we will enlarge our view of society as both face-to-face and digitalized life and discuss how the concept of social resilience should take into account the interplay of these spaces. We will then zoom into the digital world and discuss conditions for social resilience from the perspective of universal design to counter possible inequalities. We also focus on one especially crucial aspect: education. Examples will show how developing collaborative agency can take place from an early life stage by creating stimulating learning environments, face to face or digital.

1.2 Collaborative agency as the condition of resilient practice

In this section, we discuss what collaboration is and why it is an essential condition for developing social resilience. From the viewpoint of collaboration, resilience is not individual at all, resilience is a collaborative phenomenon.

Collaboration, as we perceive it (Andriessen & Baker, 2020), is a specific way of working together and relating to others. First, collaborating means developing, in an equal and mutually respectful way, a shared view of a situation in which people find themselves. "Equal" does not mean "the same," because people are obviously different in many respects, and some of them, such as complementary viewpoints or skills, may be very beneficial to collaboration. But it does mean that people should be treated as equals, in terms of mutual respect for their views, and that they should be genuinely free to say what they think about the joint task at hand. That freedom has limits, of course—those of mutual respect. Second, collaboration involves people working together to create shared meanings about what they are doing. Meanings can concern shared ideas about how to organize a fun end-of-the-year party or a new chair to be designed, a Lego sculpture, a math problem, even how to live together in shared housing. When they collaborate, people do not just all work separately to achieve a shared objective; they do that, yes, but they discuss while doing so to check whether they all have more or less the same idea about what they are doing, and even their underlying values. Third, collaboration means respecting people and not only freely cocreating shared ideas about a joint project: Collaboration is a way of working together which involves mutual respect for the dignity, the feelings, the person, the lifeworld, and the experience of all concerned.

How does collaboration, or the lack of it, relate to the notion of social resilience? We try to illustrate this through an example from the domain of cybersecurity, which is about protecting computers and computer networks from harm inflicted by hostile software, developed with malicious intention by various groups of people. Resilience in this example can be defined as the agency of individuals, as well as of their organization, to handle such cyberattacks effectively.

Example 1. Context: mid-sized municipality of a European country, such as the Netherlands. Our subject is an employee K working in the mayor's department of this municipality. One of the tasks of this employee is to read all mail the mayor receives, which may comprise several hundred emails every day. In spite of strict regulations, K opened an infected email. This message had no subject, which could have caused suspicion. After K opened the email, files in each directory on his computer he accessed became inaccessible. The IT department was notified. They discovered a type of ransomware which had encrypted all the files in the directories K had accessed. The computer was restored in the IT department, where all files not yet accessed were retrieved, the system was reinstalled, and the intact files were restored. The infected files could not be retrieved. K was initially terrified, fearing the computer was lost, but now sees this as a lesson learned, to more strictly apply the regulations in such cases. From the viewpoint of the individual, K is on his own: He makes the mistake of not attending to regulations. If this happens more frequently, sanctions will be his sad destiny. For the organization the same may happen again to another employee as nothing is shared. At most, regulations will be reinforced for all. The reason why nothing is shared (with other employees, departments, municipalities) has to do with shame (for the individual), damage of reputation (for the IT department), and loss of trust in the safety of the municipality's services. Moreover, we noted a tendency in this organization to blame the IT department for a lack of protection and slowness in service. What we see develop are defensive tendencies to counter such allegations and not the required social resilience to meet cybersecurity challenges on a group level, for the whole organization.

What would a more collaborative approach to this situation look like at the organizational level? Let us look at the three dimensions of collaboration we discussed earlier. In terms of equality, it seems all members of the organization are in the same boat: Attacks from outside (as well as from inside) can happen to all, and therefore this is a joint problem. If cybersecurity is seen as a joint problem, then it is not enough to help the individual and show him a rule book. Instead, the organization needs to better understand the options for its own resilience. In this case, we suggest two strong collaborative implications for such resilience: awareness and sharing. Awareness means all stakeholders undertake (joint) efforts to understand the challenges posed by cybersecurity. Sharing means to be open to others about incidents and how they were met and resolved. Concerning mutual respect, the third criterion, means not blaming anyone or any group in particular but jointly acting to meet cybersecurity challenges. The collaborative condition of resilience implies that all stakeholders develop the agency to act according to these principles: understand cybersecurity as a joint problem, respect the roles and positions of others, uphold transparency about incidents and solutions, and share as much information so as to maintain collective awareness. At the specific level of a cybersecurity incident, we can imagine users collaborating on (1) detection and understanding of an issue, (2) deciding what to do, and (3) informing other users of what happened. This could be done formally, by following regulations, informally, or both. What we call collaboration here is, therefore, more than discussion. The main reason we focus on collaboration is because we insist on considering the other in this activity as equals, as being in the same boat, and on envisaging a joint goal of combatting cybercrime. From this, we can see how being able to collaborate on this challenging topic could result in developing social resilience as a group or organization. This is an attitude as much as it is a vision of how to act.

Leading policymakers, educators, and other stakeholders in society agree that it is important that children in education develop skills which empower them to understand, deal with, and generate changes in society (e.g., Eurydice, 2012; The Budapest agenda, 2015; The Oslo agenda, 2006). Education is central to economic and societal progress and pivotal to the socioeconomic outcomes and life chances of individuals (Goldin & Katz, 2007; Van De Werfhorst & Van Tubergen, 2007). Many researchers label education as academic, creativity-killing (Robinson, 2006), transmission-based (e.g., Andriessen et al., 2003), founded on an inappropriate container metaphor for learning (Bereiter, 2002), detached from the real world (Resnick, 1987), and, most importantly, reluctant to change, although there are differences between nations in this respect (OED, 2014a).

The following example shows collaboration in the context of elementary school education to prove its value even in an early life stage.

Example 2. Context: a medium-sized elementary school in the Netherlands. It is an ordinary start of the school day in a preschool classroom for the 4- and 5-year olds. But not completely. It is December and the teacher is decorating the classroom with the help of the children and some parents—to create a Christmas atmosphere. There is a Christmas tree in the center of the room on a tabletop, and there the teacher is busy with some decorations, handing over items to several children to put on the tree. The room is noisy. Kids and their parents are entering, some boys are loudly moving around, some parents are chatting together while other parents are seated with their small child on their laps, looking around or reading a book together. The parents will leave when the music will sound, which will be in 5 minutes. The teacher sees two girls next to the Christmas tree following his actions with intense interest. The tabletop is covered with colored paper, but the paper is not yet properly arranged and attached. The teacher hands the two girls a roll of tape, asking them to together fix the colored paper to the table, then he turns around in another direction. The kids look at the tape, and at each other, clearly not knowing what to do, how to approach this task. A mother interprets their signs of initial bewilderment as fixed and takes the tape from their hands, cuts off three pieces, and fixes the paper to the table. While she does this, expertly, the two kids leave in another direction.

The teacher provides the children the opportunity of helping him, but by not explaining to them how to act, he invites the children to discover for themselves how to act together. This is a structure affording children to learn from experience. It includes direct instruction but leaves room for the development of collaborative agency. However, by taking over for the kids, the parent causes the children to lose interest instead of learning something. The opening created by the teacher is gone. Development of collaborative agency/collective action starts by teachers structuring and children acting in seemingly natural situations, the potential to act and create is the upshot of what happens in these situations. Learning this involves all participants in the educational community, including those at home. Some teachers understand this, others need more feedback, and the same applies to parents. There are plenty of learning opportunities to grasp, but they can be easily lost. A more systematic pedagogical approach to providing structure for the development of collaborative agency/collective action will benefit all members of the community. The abovementioned example shows how collaborative agency can develop within the education structural frame. Providing opportunities for children to collaborate on open tasks could be one way of engaging them in the process of developing agency, that is, how to act together, share goals, and take each other into account as equals.

Even though it is often economic adversity that is being pronounced in the sociological analysis of resilience (Revilla et al., 2018), the mere focus on economic development would not be analytically relevant in understanding social resilience. Hall and Lamont (2013) emphasize the power of meaning structures which both cocreate crisis and assist in the reestablishment of well-being. Keeping in mind the group notion of social resilience, it is again the social group within which new meanings should appear and support resilient action. Thus the focus on group is empowering, because the power of meaning systems can only be (dis)approved within social interactions, and, as we showed earlier, collaboration is a condition for the development of social resilience. In contemporary times, it is no longer possible to single out digitalized group dynamics when analyzing social meanings. Therefore in the sociological analysis of resilience, it is good to keep both online and offline spheres of social dialog in the scope. Unless shared meanings among social groups are being restored with digital and spatial solutions, they will not work toward resilient actions. It may happen as well that while looking at the structures of shared meanings which work toward social resilience, new group dynamics and compositions will be uncovered, and thus the resilient action will be empowered. However, without keeping offline development cautiously in scope, it may happen that existing social dysfunctions and tendencies to increase social inequalities only grow with the introduction of new technologies in searching for innovative solutions. The research on how vulnerable social groups use new technologies and cocreate digital solutions to contemporary social challenges brings essential inputs to the search for social innovations which assist and empower resilient actions, as well as among vulnerable populations of ethnic-minority background or low capacity digital devices use (Ereky-Stevens et al., 2019).

To explain the interplay between the technological and human layers of resilient practices, let us present two examples of public transport ticket selling processes by vending machines in two European cities, London Victoria station in July 2018 and Lisbon Airport in October 2019. Both places are frequent transportation hubs with plenty of foreign incomers included. Language and cultural barriers are therefore involved in the practice of purchasing the tickets. At the same time, these are crowded places, with numerous rapid and stressful interactions.

Example 3. Context: London Victoria station. I am approaching the vending machine to return the Oyster card, but I am not sure how and if I will get my money back, and also how I would then continue to the train to the airport. There are several vending machines and two uniformed Transport for London employees. The employees are quickly serving the costumers approaching them, sometimes with short questions or navigation requests, sometimes with the need for assistance. The lady I approach goes with me through the entire procedure. I now feel safe, and I also think she saved my time by avoiding all the possible mistakes. I am leaving the place fast. It still looks very messy from a distance, but inside there is an invisible order of social and technical interactions serving the transportation purpose and ad hoc needs of the passengers.

Example 4. Context: Lisbon Airport metro foyer. I am entering the metro foyer. There is a long cue for the vending machines. As I queue, I see some people exiting the metro waving with their cards offering to give them away. Sometimes a person steps out of the cue and takes the card. I observe the practice for a while and when someone waves toward me, I take the card, leave the cue, and try to put it into the terminal. As I go toward it, I can see a poster from Metropolitano de Lisboa written in English warning not to take cards from other passengers. I have conflicting feelings because the person I took it from did not look like a criminal, and I do not consider myself one either. It seems it was another traveler leaving Lisbon. I enter the public transport area with mixed feelings.

In both cases, there are passengers with limited knowledge of the given system involved. In the case of London, the presence of personnel by vending machines has both a channeling and controlling function and also works to smoothen the operation of services. In the case of Lisbon, the English poster manifests the symbolic message of being aware of non-familiar users, but the notification is not intended for the diverse users. On the contrary, the message is neither easy to find nor making use of the vending machines flexible. Another technological solution that would be more inclusive is paying for the transport via smartphone, but this is potentially exclusory of low-threshold users. To summarize, even in smart cities, the physical presence of human assistants may still be the most efficient way of providing services to a diverse target group, and therefore supporting collaborative agency without making actors feel guilty of collaboration which positions them against the system.

1.3 Sensitive design and digital tools for supporting social resilience

This section concentrates on the way environments, especially digital ones, impact the dynamics of equality and shared meaning-making and problem-solving—how they can stimulate or inhibit social resilient practices in participants. A latent risk in the design of digital tools and services is renewed perpetration of segregation and exclusion forms which come from long-established, hard-to-defeat prejudices or just from a lack of attention toward less visible forms of inequality.

As Robinson and Myck-Wayne (2016) argue, "Digital inequality deserves a place alongside more traditional forms of inequality in the twenty-first century pantheon of inequalities." To shed light on this specific problem, two types of relationships between digital exclusion and design will be discussed. First, we will take a look at how design can contribute to social exclusion and even segregation. Second, we will examine how nonstandard users tend to be left out of the usage of contemporary digital tools. Furthermore, some suggestions related to universal design will be formulated.

First of all, we will consider forms and actions of digital exclusion that are more or less intentional or that exist "by design." In the last few years the topic of "exclusion by design" has emerged, for instance, as a critique of intelligent systems based on decision algorithms (Eubanks, 2018; Park & Humphry, 2019). The introduction of automated systems reinforces punitive policies of existing service regimes intentionally, or, as we have already stated, by design. Processes of data collection, data mining, and profiling are then perpetrating existing digital inequalities by creating new scenarios where being an exception is not considered. Decisions are driven to fit a statistical model; people from specific ethnic groups, with a certain income, or passing through specific life situations (such as losing a job or having a disease) are more likely to suffer from consequences related, for instance, to insurance plans or qualifications for benefits. As Cobo (2020) puts it, "Algorithms are often designed with a reductionist view that simplifies or ignores contexts (where information is generated or transformed), thus standardizing a set of data to fit other systems."

A second way in which digital exclusion and design are connected relates to (mostly) unintentional processes. It is exclusion created by design choices driven by one simple idea: that "average users" are capable, enabled, entitled, and allowed to do certain things, and that if they are not, they are "special users." This view has been criticized mostly in the usability and accessibility fields by highlighting the negative effects it can have on access to the built environment (Mace, 1985), educational content (Rose & Meyer, 2002; Scott et al., 2003), and businesses (Keates, 2006). The attention is mainly on people with disabilities, but it can be scaled up to include all the variability typical of human beings. Keeping a close eye on the topic of access to digital resources, those who do not have "primary access" to them (i.e., as a consequence of digital divide or because they cannot afford a personal device) are the first to have reduced access to information and services. They often develop negative attitudes toward technology, further deterring them from Internet adoption (Mesch & Talmud, 2011). The Pew Research Center "Internet/Broadband Fact Sheet" (Pew Research Centre, 2020) shows how, in the United States in 2019, there are still visible differences related to age, ethnic group, level of income, and education level. Gender differences in access are reduced in that context, but, in many other parts of the world, they are still a big issue (Anandhita & Ariansyah, 2018; Badran, 2018), and even improved access does not automatically mean more empowerment (Bailur et al., 2018). Simply considering a tool or a digital environment accessible because it is available to "most" users, does not mean it is designed to be used by all. It is a matter of prerequisites, sometimes difficult to discover in a complex world where inequalities can manifest themselves at different levels.

To elaborate the notions of sensitive design and how they could be used to promote collaboration and give space for social resilience, we will consider the seven principles of universal design as they are presented in the Centre for Excellence in Universal Design formulation (CEUD, 1997). We will also provide examples of how these principles should be read to avoid the two forms of digital exclusion discussed previously as well as how they can be useful for the stimulation of social resilience. For each principle, we will propose a formulation that is more oriented toward the types of inequalities which may eliminate the potential for collaborative agency.

1.3.1 Universal design principles

1.3.1.1 Equitable use

> *The design is useful and marketable to people with diverse abilities.*
>
> **CEUD (1997)**

This must include people who have diverse levels of education and digital literacy and who have variable access to digital devices and services. Testing on a variety of devices, from low- to high end, should be a priority for all digital platforms.

1.3.1.2 Flexibility in use

> *The design accommodates a wide range of individual preferences and abilities.*
>
> **CEUD (1997)**

The domain of "preferences and abilities" can be extended to the use of low-end, outdated, or slow devices or to a lack of reliable Internet connectivity. This moves the focus slightly away from the actual user but should help in taking into consideration his or her actual level of performance, considered the result of preferences and abilities situated within a context.

1.3.1.3 Simple and intuitive use

> *Use of the design is easy to understand, regardless of the user's experience, knowledge, language skills, or current concentration level.*
>
> **CEUD (1997)**

The concept of "easy to understand" is often directly linked to the level of scholarship when readability and comprehensibility assessments are performed. This is not only particularly important according to this principle but is also something that can be assessed in a target population. A key design principle to take into consideration is that of "consistency": interfaces which look familiar and use conventional icons and a shared, familiar visual language are more likely to be usable by a broader public.

1.3.1.4 Perceptible information

The design communicates necessary information effectively to the user, regardless of ambient conditions or the user's sensory abilities.

CEUD (1997)

Besides the need for different forms of communication (i.e., visual and auditory) to maximize the perceptibility of information, it is important to consider that differences in perception abilities can be caused by different traits, may they be social or cultural. This does not in any way mean that people from different cultural backgrounds will be unable to process a piece of information online; nevertheless, it is not a given that they will treat it with the adequate degree of attention.

1.3.1.5 Tolerance for error

The design minimizes hazards and the adverse consequences of accidental or unintended actions.

CEUD (1997)

Being sensitive to disadvantaged contexts means to prevent all kind of physical or psychological damage, as well as potentially upsetting situations. This includes checking whether a specific idea, word, or picture (i.e., published in a social media campaign) could have a particular meaning in other cultures, or if there is some kind of offense which could be taken from content addressing a diverse audience.

1.3.1.6 Low physical effort

The design can be used efficiently and comfortably and with a minimum of fatigue.

CEUD (1997)

Reducing the required efforts in a digital environment mostly means minimizing repetitive actions and allowing use from a neutral, natural body position. To address this point, it is today particularly interesting to leverage the possibilities opened by new kinds of digital interfaces, such as speech to text, chatbots, and virtual assistants.

1.3.1.7 Size and space for approach and use

Appropriate size and space is provided for approach, reach, manipulation, and use regardless of user's body size, posture, or mobility.

CEUD (1997)

This last principle gives great importance to the context where interactions happen. For a digital tool to be compliant with this recommendation, it is important that it is compatible with other tools and platforms already in use and that it does not require a large amount of resources.

These principles, originally formulated as a recommendation to architects and product designers, cover various aspects of design and implementation processes. The overall idea is to prevent situations which could be problematic by keeping an eye on different dimensions of the design process itself. This view of the good designer as the mind behind effective projects, useful devices, and a welcoming environment might overlook one other dimension which is crucial to the development of resilient practices: collaboration.

Through it, and especially by fostering collaborative agency, a system—be it a public space, a piece of learning content, or even a meeting between human beings mediated by some sort of digital technology—can improve its accessibility and inclusiveness.

Example 5. Context: university virtual classroom. At the beginning of an online class, a university professor tries to connect to the university e-learning platform and finds it offline: Too many students are trying to connect, since it is the start of the semester, causing a system overload in a university which had promptly switched to online teaching as a response to the COVID-19 emergency. The teacher finds himself online with just a few students who were able to connect to an external videoconference platform just some seconds before the learning management system collapsed. By communicating the direct link of the videoconference session to these few "survivors," the teacher asks them to pass the information on to other students in the same course by other means of communication. This translates, a few minutes later, into more than 100 students connected to the platform: They received the invitation link mostly via WhatsApp from other students who knew them from other courses or workshops.

Example 6. Context: hearing disabilities and online chat participants. During the COVID-19 emergency, many people with hearing loss or deafness were forced by the situation, together with people with perfect hearing condition, to spend much more time online than usual. This time involved, for most knowledge workers, participation in (a lot of) videoconferences. The inaccessible nature of such events was already visible: Only a few platforms had proper support for sign language interpretation and only a few more had automatic captioning. This problem, normally quite easy to avoid by using other communication means, was brought rapidly into the center of the work life of many people sharing that specific condition. An easy way to solve the issue is using online services, such as those of webcaptioner.com, to produce live subtitling—a presenter could share his or her screen view to accommodate the needs of disabled participants. But without this form of collaboration, it is not easy (but neither impossible) to provide captioning for the same speech with the same accuracy.

Example 7. Context: EUvsVirus Hackathon. Information and Communications Technology (ICT) and human capital is being used to tackle the sudden global crisis and the jeopardy of collaborative values. It is the second half of April 2020 and most European countries are now experiencing a lockdown due to the COVID-19 epidemic situation. I have received an invitation to take part as a mentor in a hackathon regarding innovative projects which can beat COVID-19 and the secondary consequences straining economies, social relations, and infrastructure. The invitation comes by email through a professional association of which I am a member. Since it is organized by the European Commission, the key scientific research partner in the European Union, I decide to take part. Within days a typhoon of social media and digital management apps passes through my work desk. I first register with the Slack interface, then watch the opening ceremony via Facebook, search through competing projects via Devpost, and register my competencies at Airtable. I am being encouraged to get the Hackathon badge from LinkedIn. When the event launches, I am addressed via Slack by diverse teams; I suggest some reading, terminology, consult on proceedings, and provide a user test for the app being designed. I am trying to get another expert to one of the national teams via WhatsApp—we found each other by posting a national flag to chat on Slack during the opening ceremony. Some days later, after the results are announced, we watch the live performance of the song together. The event attracted over 30,000 people from across Europe. Taking part made me feel "old normal"—interconnected, cosmopolitan, innovative. Nevertheless, I also felt the tendency to single out only those who are fluent in English, strong in ICT use, and have sufficient Internet connectivity; so the entire event felt inclusive but also exclusive to a degree.

The examples given above highlight how accessibility problems can arise anywhere and anytime and that a prompt response from users can be a way to solve issues efficiently. The collaboration can foster the use of digital management tools and social media when facing an unprecedented crisis, nevertheless, the existing social barriers remain in play.

1.4 Conclusion

Social resilience can have many forms that may be partly designed, facilitated, and permitted. What these forms have in common is its capacity to create innovative solutions for unforeseen challenges while incorporating the existing diverse inner resources of actors taking part in looking for solutions. Technologies, on one hand, have the power to amplify existing inequalities, but they also have the potential to open new spaces to tackle adversity and may create opportunities to redefine collaborative action so it becomes more resilient.

In the field of architecture and urbanism, it became a habit to formulate key strategic principles of resilience (see part 5.2 in Colucci, 2012). What follows is a summary of these principles of supporting social resilience:

1. Keep in mind group dynamics and interactions. Social resilience is always a matter of multiple actors; it cannot be reached or regarded on an individual basis.
2. Collaborative action makes social resilience possible. Sharing views and proceedings, searching for common meanings of actions taken, and respecting every participant involved are the conditions of collaborative action.
3. Collaborative agency is key. Reestablishing innovative structures and new meanings is where social resilience is maintained.
4. Technological support of innovative solutions should be designed sensitively. The goal is to include as many diverse users as possible, not to design "one-size-fits-all" solutions.
5. Human and technological actions coexist. Both are integral parts of innovative solutions for unforeseen crises and the persistent adversities of contemporary social and environmental conditions. The interaction of these two types of actions should be closely and continuously evaluated to reach social resilience.

The English word resilience has only gained special conceptual meaning in recent decades. Its original meaning dates back to the 16th century and denotes breaking an oral agreement, which is far from the usual contemporary meaning (OED, 2014b, in Revilla et al., 2018, p. 89). Nevertheless, one commonality is found here. The term resilience is, and always has been, a dual term, its meaning only able to be explained in close relationship to a breakdown, a crisis. It appears in the complementary context of various terms that identically denominate discomfort, difficulty, and resistance. It is either adversity as a concentrated form of social suffering (Revilla et al., 2018), "austerity measures under neoliberal economics" (Welch & Scott, 2016 in Gallagher et al., 2019, p. 189), or respectively, upheaval. To an extent, the term legitimizes and states the unsettled state of the world we live in, and it provides an optimistic view of the capacity to deal with the difficulties of our time. After the age of modern normativity and postmodern inclusivity, these are times of accepted dilapidation and a neither optimistic nor dystopian but rather collaborative and resilient outlook moving forwards.

References

Anandhita, V.H., & Ariansyah, K. (2018). Gender inequality on the internet access and use in Indonesia: Evidence and implications, in Proceeding *of the international conference* on ICT for *rural development*: Rural development through ICT: Concept, design, and implication, IC-ICTRuDEv 2018, pp. 142–147. <https://doi.org/10.1109/ICICTR.2018.8706856>

Andriessen, J., & Baker, M. (2020). *On collaboration: Personal, educational and societal arenas*. Leiden: Sense/Brill.

Andriessen, J., Baker, M., & Suthers, D. (2003). Argumentation, computer support, and the educational context of confronting cognitions. In J. Andriessen, M. Baker, & D. Suthers (Eds.), *Arguing to learn* (pp. 1–25). Netherlands: Springer. Available from https://doi.org/10.1007/978-94-017-0781-7_1.

Badran, M. F. (2018). *Bridging the gender digital divide in the Arab region. The future of work in the Global South* (pp. 34–39). IDRC.

Bailur, S., Masiero, S., & Tacchi, J. (2018). Gender, mobile and development: The theory and practice of empowerment. *Information technologies and international development, 14* (2018), 96–104.

Bereiter, C. (2002). *Education and mind in the knowledge age*. Routledge.

CEUD. (1997). What is universal design? Available from http://universaldesign.ie/What-is-Universal-Design/The-7-Principles/.

Cobo, C. (2020). *I accept the terms and conditions: Uses and abuses of digital technologies*. Montevideo: IDRC and Ceibal Foundation.

Colucci, A. (2012). Towards resilient cities: Comparing approaches/strategies. *TeMA: Journal of Land Use, Mobility and Environment, 2*(2012), 101–116. Available from https://doi.org/10.6092/1970-9870/921, ISSN 1970-9870.

Dreier, O. (1999). Personal trajectories of participation across contexts of social practice. *Outlines. Critical Practice Studies, 1*(1), 5–32.

Eubanks, V. (2018). *Automating inequality: How high-tech tools profile, police, and punish the Poor*. New York, NY: St. Martin's Press.

Eurydice. (2012). *Developing key competences: Challenges and opportunities for policy*. Eurydice.

Gallagher, C. W., Minter, D., & Stenberg, S. J. (2019). Introduction. *Pedagogy: Critical Approaches to Teaching Literature, Language, Composition, and Culture, 19*(2). Available from https://doi.org/10.1215/15314200-7295849.

Giddens, A. (1984). *The constitution of society*. Cambridge: Polity Press.

Goldin, C., & Katz, L. F. (2007). *The race between education and technology: The evolution of US educational wage differentials, 1890 to 2005*. National Bureau of Economic Research. (No. w12984).

Hall, P. A., & Lamont, M. (2013). *Social resilience in the neo-liberal era*. Cambridge: Cambridge University Press.

ICT-facilitated parent interventions to support families with young children in linguistically diverse contexts. In K. Ereky-Stevens, M. Trauernicht, A. S. Sarcinelli, & K. Sidiropulu-Janků (Eds.), *Design and implementation, using the ISOTIS Virtual Learning Environment*. ISOTIS. Available from https://staging-isotis-pw.framework.pt/site/assets/files/1480/20191201_isotis_wp3_vle_report.pdf.

Keates, S. (2006). *Designing for accessibility: A business guide to countering design exclusion. Human factors and ergonomics*. Boca Ratin: CRC Press.

Mace, R. (1985). *Universal design, barrier free environments for everyone*. Los Angeles: Designers West.

Mesch, G. S., & Talmud, I. (2011). Ethnic differences in internet access: The role of occupation and exposure. *Information communication and society, 14*(4), 445–471. Available from https://doi.org/10.1080/1369118X.2011.562218.

OED. (2014a). Measuring Innovation in Education. *A new perspective*. Available from https://doi.org/10.1787/9789264215696-en.

OED. (2014b). *Oxford English Dictionary*. Oxford: Oxford University Press.

Park, S., & Humphry, J. (2019). Exclusion by design: Intersections of social, digital and data exclusion. *Information communication and society, 22*(7), 934–953. Available from https://doi.org/10.1080/1369118X.2019.1606266.

Pew Research Centre. (2020). Available from https://www.pewresearch.org/internet/fact-sheet/internet-broadband/.

Resnick, L. B. (1987). The 1987 presidential address: Learning in school and out. *Educational researcher, 16*(9), 13–20.

Revilla, J. C., Martín, P., & De Castro, C. (2018). The reconstruction of resilience as a social and collective phenomenon: poverty and coping capacity during the economic crisis. *European Societies, 20*(1), 89–110. Available from https://doi.org/10.1080/14616696.2017.1346195, ISSN 1461-6696.

Robinson, K. (2006). Do schools kill creativity? TED talk. Available from https://www.ted.com/talks/sir_ken_robinson_do_schools_kill_creativity/transcript.

Robinson, S., & Myck-Wayne, J. (2016). A teacher training model for improving social facilitation in the inclusive program. *Young exceptional children, 19*(1), 16–26. Available from https://doi.org/10.1177/1096250615586028.

Rose, D., & Meyer, A. (2002). *Teaching Every Student in the Digital Age: Universal Design for Learning*. Alexandria, VA: ASCD.

Scott, S.S., McGuire, J.M., & Shaw, S.F. (2003). Universal Design for Instruction: A New Paradigm for Adult Instruction in Postsecondary Education. *Remedial and Special Education*. v24 n6, 369-79

The Budapest agenda (2015). Available from http://ec.europa.eu/enterprise/policies/sme/promoting-entrepreneurship/education-training-entrepreneurship/teacher-education-entrepreneurship/index_en.htm.

The Oslo agenda (2006). Available from http://www.gvpartners.com/web/pdf/EC_Oslo_Agenda.pdf.

Van De Werfhorst, H. G., & Van Tubergen, F. (2007). Ethnicity, schooling, and merit in the Netherlands. *Ethnicities*, 7(3), 416–444. Available from https://doi.org/10.1177/1468796807080236.

Welch, N., & Scott, T. (Eds.), (2016). *Composition in the age of austerity*. Logan: Utah State University Press.

CHAPTER 2

Investment, risk, and vulnerability: perspectives in adaptation and risk finance

Luca Cetara
Eurac Research

2.1 Introduction and methods

Funding is a requirement for reaching sustainability goals. This fact has been recognized at large (Schoenmaker, 2017). The potential role of sustainable finance has become a major topic worldwide and particularly in Europe in application of the EU action plan on sustainable finance (EC, 2018). The EU has committed to channel private financial flows toward investments that support the target set by the Paris Agreement of a neutral-carbon economy by 2050, and the United Nations Sustainable Development Goals. In general, the issue of mobilizing finance for sustainability received a growing attention over the last few years also by the private financial sector (Levenson Kehoane & Madsbjerg, 2016). Sustainable finance has been defined as referring to the integration of environmental, social, and governance considerations in processes of financial decision-making (Archer, 2019). A large part of the debate and contents on sustainable finance refers to the topic of climate finance. However, most of the financial institutions have concentrated their efforts in addressing the mitigation side of climate change finance dealing with interventions aimed at reducing CO_2 and other GHG emissions from the world economies. Only a limited but growing body of research has addressed the side of adaptation to climate change and the possibility to set up appropriate instruments traded or offered by financial institutions and markets in support of investment in more resilient and adapted economies and societies.

Here, we focus on the narrower field of finance for adaptation to climate change, by addressing the issue of how a finance-oriented interpretation of the vulnerability and risk functions as used in the domains of climate change adaptation and disaster risk reduction (DRR) can provide insights and potential room for introducing financial instruments and driving investment to

[ECO]systems of Resilience Practices
DOI: https://doi.org/10.1016/B978-0-12-819198-9.00006-5

address and support adaptation, and risk and damage mitigation, in line with the demands for a more resilient society.

By addressing the topic of using investment for supporting vulnerability and risk minimization objectives, we aim at:

1. proposing an identification of the components of the vulnerability, risk, and economic damage conceptualizations that adaptation-oriented finance can address and explaining how this might happen;
2. explaining how the variables of the vulnerability, risk, and damage function usually applied to climate change adaptation and DRR can provide information in support of investment decisions in line with the generally accepted adaptation and DRR policy targets; and
3. identifying a relationship between adaptation finance and sustainable finance as the latter has been set in literature.

The chapter is based on qualitative and literature research in two main domains. The first one refers to resilience, particularly applied to the case of climate change adaptation, and includes sources in the fields of climate change adaptation, vulnerability assessment, DRR, and quantification of economic damage from natural hazards.

The second one refers to sustainable or green finance and, to a minor extent, to climate finance. No literature on public finance, macroeconomics, and environmental expenditure at different levels has been surveyed, since we aimed at identifying general frameworks for sustainable financial decision-making mainly for the private sector and by institutional investors. Occasionally, considerations concerning the social effects of private decisions in the field of climate actions and finance have been recalled.

We provide a concise analysis of the concepts and simple expressions of vulnerability, risk, and damage and focus on the variables susceptible to be influenced by financial decision-making. The descriptions mainly focus on the relevant aspects of the concepts for the financial sector. We often supply narrative descriptions of concepts, their relationships, and possible consequences of alternative decisions and policies.

At first, we attempt to identify the variables on the basis of the concepts of vulnerability, risk, and damage. Then we try a critical analysis of the variables susceptible to be influenced by investment decisions by the financial sector and an assessment of the consistency of the likely resulting behaviors to the general targets of public adaptation policy. Based on literature, we aim at supplying an overview on the reasons for a possible motivated increased involvement of the financial sector in the implementation of consistent adaptation actions in line with the targets of public policies for adaptation and risk mitigation.

2.2 Addressing resilience in finance: the basic dimensions

In climate change literature, two terms applied to the target of an impact express the potential (negative) effects of climate change on ecosystems, infrastructure, economic sectors, social groups, communities, regions, or socioecological systems in general, that is, vulnerability and risk.

2.2.1 Vulnerability

According to IPCC (2014), vulnerability to climate change is the propensity or predisposition of a system or a target to be adversely affected. Vulnerability encompasses a variety of concepts, including sensitivity or susceptibility to harm (sometimes fragility), and lack of capacity to cope and adapt. In its simplest version, vulnerability is a function of the character, magnitude, and rate of climate change and variation to which a system is exposed, its sensitivity, and its adaptive capacity.

Therefore vulnerability (V) can be expressed as depending on three variables: exposure (e), sensitivity (s), and adaptive capacity (ac).

Exposure (e) refers to the presence of people, livelihoods, species or ecosystems, environmental functions, services, and resources, infrastructure, or economic, social, or cultural assets in places and settings that could be adversely affected.

Sensitivity (s) refers to the degree to which a system or species is affected, either adversely or beneficially, by climate variability or change. The effect may be direct (e.g., a change in crop yield in response to a change in the mean, range, or variability of temperature) or indirect (e.g., damages caused by an increase in the frequency of flooding).

Adaptive capacity (ac) refers to the ability of a system to adjust to climate change (including climate variability and extremes) to moderate potential damages, to take advantages of opportunities, or to cope with the consequences (IPCC, 2014).

If we consider the goal of adaptation policies at large the one of adjusting human and natural systems to actual or expected climate and its effects by moderating or avoiding harm or exploiting benefits (IPCC, 2014), then we can assume as socially desirable the minimization of the vulnerability to climate change impacts of a system or a more specified target.

In simple analytical terms a vulnerability function depends in a direct proportion on exposure, sensitivity, and in a reverse proportion on adaptive capacity. It can be expressed by the following relationship:

$$V = f(e, s, ac) \tag{2.1}$$

A particular case of this function is the one where vulnerability (V) depends on the concepts of *potential impact* (pi) and *adaptive capacity* (ac), where:

$$pi = f(e, s) \tag{2.2}$$

which can be expressed as:

$$pi = e \times s \tag{2.3}$$

Thus a *vulnerability function* (V) can also be written as:

$$V(pi, ac) = pi - ac \tag{2.4}$$

Function (2.4) allows separating the adaptive capacity of a system from the other two variables on which vulnerability depends, referred to as *potential impact.* This might be useful for analyzing the potential contribution of finance and investment to minimizing vulnerability through increasing the adaptive capacity of the target of an investment.

2.2.2 Risk

We might use a risk function to insert the vulnerability measure within a framework that takes into account also the physical dimension of the impact depending on climate forcing or physical phenomena.

In this way the *objective function* becomes more like:

$$R = f(H, V, e) \tag{2.5}$$

where $V = f\ (s, ac)$

Therefore the consequences of an event depend on all the mentioned variables:

$$C = f(H, V, e) \tag{2.6}$$

This can also be written as follows:

$$C = f(H, s, ac, e) \tag{2.7}$$

Additionally, there is an issue of likelihood or probability of the event, where:

$$P = f(H, V) \tag{2.8}$$

This can also be written as:

$$P = f(H, s, ac) \tag{2.9}$$

This means that hazard and vulnerability can both contribute to the "probability" of an event: the former to the likelihood of the physical event and the latter to the likelihood of the consequences resulting from the event.

In its simplest expression, risk (R) directly depends upon the probability of an event (p) and its consequences (c) expressed in any form (here the issue with metrics comes up):

$$R = p * c \tag{2.10}$$

We find that climate change risk (R) depends on the vulnerability function as discussed in Eq. (2.4) according to a direct relationship, so that $f_v'(R) > 0$. It also depends on a hazard factor (H) that is fundamentally physical[1], as in Eq. (2.5) and is subject to the probability function (2.9) depending on the type of hazard and the vulnerability at play in the case.

The advantage of the notation used in Eq. (2.5) in a discussion on financial instruments basically is that it allows to fully include the conceptualization of vulnerability as presented in Eq. (2.4). The resulting notation refers to risk (R) as a driving concept that is easier to incorporate in classical financial decision-making approaches where different categories of risk are usually factored in.

2.2.3 Adaptation objectives

Based on the discussion earlier, we assume that a sound climate change adaptation policy should adopt as significant targets the following ones:

1. minimizing the value of vulnerability of the affected system or target of climate change impacts by acting on the exposure, sensitivity, and adaptive capacity variables;
2. minimizing the value of risk associated to the affected system or target of a climate hazard causing an impact on a system or target, by acting on hazard (H) when possible and on the value of vulnerability as in list (1);

[1] According to some interpreters, the hazard as it strikes a specific location can depend on some adaptation measures undertaken outside the target location (e.g., maintenance of green infrastructures outside a metropolitan city borders can mitigate the risk/magnitude of floods). This intervention could be considered as a type of adaptive capacity.

3. minimizing particularly the value of economic damage associated with a hazard or climate change impact, by addressing the probability and consequences of an event and other relevant economic variables determining the value of the exposed assets as well as the values of risk and vulnerability as in lists (1) and (2).

The achievement of these objectives can depend on changes in the variables of exposure, sensitivity, adaptive capacity, hazard, and probability that can to some extent be induced or supported by means of investment by the financial sector. Some considerations on the potential role of investment in meeting these targets will be provided next.

2.3 Financial considerations on risk and vulnerability

We wish to discuss in brief which of the terms of the vulnerability function (2.4) (namely, pi and ac) can be affected by investment decisions by financial institutions, focusing in particular on the case wherein desirable results in terms of increased resilience or adaptation are delivered as a consequence of a financial decision. We will call this case "adaptation finance" and refer to the instruments used to achieve results that increase resilience or deliver adaptation capacity (proresilience and proadaptation results) as to *adaptation financial instruments* (AFI). This discussion could provide useful hints on how the financial sector can contribute to achieve the climate policy targets being more closely related to adaptation and resilience. Additionally, it could help to identify the needed incentives or informational gaps to be filled in aiming at incentivizing larger investments consistent with adaptation or resilience objectives.

Potential impact (pi) shows a direct proportion to vulnerability changes (ΔV), thus climate adaptation finance should aim to determine a reduction on the overall amount of pi by possibly minimizing the two distinctive variables of the system affected by climate change impacts, being respectively exposure (e) and sensitivity (s).

Exposure (e) refers to economic, social, and cultural assets, people, ecosystems, and their functions that can be ideally assimilated to different types of capital, as framed by literature (Goodwin, 2003; Bourdieu, 1986). All those assets are exposed to potential climate change impacts. Since in the case of exposure, we consider a minimization target as socially desirable, all financial instruments disincentivizing the concentration of assets susceptible to be adversely struck by climate change impacts in a territory where scenarios foresee relevant effects of climate change over a certain timescale can be considered contributing to reducing vulnerability.

The principal finance-driven approaches in this direction include divestment strategies from most vulnerable assets or activities with limited residual adaptive capacity (RAC), within a territory where a climate risk assessment and scenarios are available; insurance policies disincentivizing the occurrence of situations where impacts or their associated losses are too high and not viable for marketable insurance contracts (e.g., mandatory protective measures or use of allowances); investment subject to safety or adaptive measures that reduce the exposure of the target to a specific hazard (e.g., technical adaptation measures); investment in changes in the exposed assets resulting in lower exposure and potential loss and damage (e.g., structural modifications to buildings exposed to flood hazard). Financially speaking, it is advisable to perform a viability assessment based on long-term capital budgeting, where a comparison is made between the cost of intervention (action) and the cost of no intervention (inaction) in terms of net present value, aiming to assess the contribution of resilience-building measures to create long-term value for the target of the investment (Schoenmaker & Schramade, 2019).

Sensitivity expresses the degree or magnitude of the effect suffered by a system or a component of it. Under our point of view, sensitivity (s) can be considered given at the present and evolving over time for biophysical reasons and climate change effects. We can expect a system's sensitivity to be at least partially influenced by improvements in adaptation, thus it can be seen as being inversely related to any advancement in adaptive capacity (ac) in a proportion β, where $0 < \beta < 1$. As a consequence, also due to the uneasy and diverse conceptualization of the notion of sensitivity (Füssel et al., 2006), we can assume here that finance and AFI can hardly affect the sensitivity of a system, at least in the short and medium runs, as we shortly discuss later.

Sensitivity seems to be unfit to be treated under a strictly financial dimension. Even though theoretically some actions could modify the sensitivity of a system (for both economic and ecological systems), it is possible to transfer the potential improvements to the other variables determining the vulnerability (or risk) function, namely, exposure and adaptive capacity (and sometimes the magnitude of a hazard for risk functions). Alternative readings of the concepts of risk and vulnerability make complex to frame any intervention that can determine changes in the sensitivity or fragility of a system and do not solve the intrinsic problem of measuring the cause–effect relationship between action and changes in sensitivity of a system or target. The assessment would become even more complicated if the investment decision was considered the independent variable responsible for a potential change in sensitivity, for which there are no single metrics available. This feature is typically interpreted as intrinsic to the object of analysis (often a system). However, it is theoretically possible to frame sensitivity (s) as a structural variable of a system and consider that structural modification of the system itself may induce changes in its level of sensitivity (Δs). Ideally, investment in some structural variables of a system could induce change in its sensitivity, in line with systems theory (Meadows, 1999). We will mention later the how cobenefits from other policies or investment decisions can affect the sensitivity of a system (e.g., a discussion on the effects of a shift to a circular economy on the resilience of a society can be found in Dufourmont et al., 2020). Here we assume that AFI can hardly induce any sizable change in sensitivity, at least in the short to medium run.

Adaptive capacity (ac) shows a reverse proportion to vulnerability changes (ΔV), thus climate adaptation finance should aim to determine a rise of its total amount to cause a reduction in vulnerability ($\Delta V < 0$). All financial instruments directed to increase ac can therefore be considered generating a reduction in the vulnerability (V) of a system of reference or target. AFI can help enhance adaptive capacity by increasing the preparedness, technology, approaches, techniques, defenses, and similar solutions that are available over a certain period of time (usually in the short to medium run) to counteract the negative impacts of climate change on the investment-targeted assets or systems. Adaptive capacity can often be considered a constrained or limited variable. According to IPCC (2014), "a limit is reached when adaptation efforts are unable to provide an acceptable level of security from risks to the existing objectives and values and prevent the loss of the key attributes, components, or services of ecosystems". When addressing this issue, alternative terms are used, including "thresholds" and "fixed thresholds," "tipping points," "key vulnerabilities." Basically, it is possible to identify hard (that do not change over time) and soft limits (that can change over time), often assessed in a given time. The case has been investigated for thresholds in physical systems, depending on the magnitude and rate of climate change, where exceeding those levels would lead to irreversible change or losses. If we address the issue of adaptive capacity under the point of view of economics and business studies, we are brought to recognize that some limits to adaptive capacity are linked to lack of technology, scarce innovation capacity, extreme maturity of a product or service, low opportunity costs. In this case, most of the possible adaptive capacity has already been exploited for the product, service, sector, or industry under investigation, thus we can introduce the concept of RAC. In this case, investing in adaptive capacity could show decreasing returns. The presence of decreasing returns could be an indicator of little RAC and provide a significant informative instrument to investors. In particular, the insufficient return on investment (ROI) in adaptive capacity would indicate the impossibility to reverse a high value of vulnerability resulting from an assessment, which would tend to magnify the risk function's value and discourage further investment in the activity under investigation.

Hazard (H) has a magnitude that is to be considered independent from investment. However, according to some interpreters, the impact of a hazard striking a specific location can depend on some adaptation or risk mitigation measures undertaken outside the target location (e.g., maintenance of green infrastructures or building of defensive barriers outside a metropolitan city borders can mitigate the risk or magnitude of floods in the city center). This type of interventions could be considered addressing adaptive capacity. The investor is likely to be interested in an estimate of the impact of a hazard in the specific location or on the specific target of its investment, independently from the possible reasons that reduce its magnitude. However, information on effectiveness and efficiency of protection and risk mitigation measures can make an investment safer by reducing the risk of loss and damage for the provider of finances.

Residual risk (RR) is usually referred to as the share of risk that ongoing DRR processes have not mitigated or reduced sufficiently or eliminated or prevented completely. In this chapter, it can be considered the value of the risk left after efforts—supported by investment—to increase the adaptive capacity and reduce the exposure of the targeted system have been implemented. It can be a partial indicator of the risk associated with a specific hazard that cannot be further reduced efficiently by means of actions on exposure and adaptive capacity. RR cannot be avoided under any credible scenario. It affects human settlements and natural systems and may result in economic and noneconomic losses and damages. However, the presence of some framework conditions may allow using AFI to manage the transfer of RR by spreading it across different stakeholders, regions, and time or even the reintroduction of it in the risk management cycle. Under the financial point of view, this would lead financial decision makers to reconsider investment decisions after RR has been reduced by transfer or management. A low degree of RR and the presence of some room for its spreading can therefore be useful indicators for the financial sector.

Economic disaster risk, loss, and damage are briefly presented next.

Damage is usually defined as the total or partial destruction of physical assets and infrastructure in disaster-affected areas. It can be expressed as replacement and/or repair costs. Loss refers to the changes in economic flows occurring as a result of a disaster and has a more genuine macroeconomic nature.

Damage is dependent on the type of hazard and impacts expected as well as on the vulnerability assessed for a given target, the value of which is embedded in exposure. Damages are relatively easy to assess ex post; however, physical impacts are more difficult to translate into expected future risks. This limitation has been considered, inducing risks of mispricing climate risks and opportunities deriving to financial actors from possible investments, strategies, and portfolios (GCA Global Commission for Adaptation & UNEP, 2019). It is therefore a factor of potential reduction of proresilience investment.

Economic disaster risk may be defined as a probability distribution indicating potential economic damage costs and associated return periods. Certainly, this distribution may depend on several further variables than the one directly used in vulnerability and risk assessment; however, hazards and climate impact scenarios on specific areas and targets (through the targets' characterization in terms of sensitivity and adaptive capacity) can help determine the probability of an event, its physical impacts and return periods that significantly participate in causing the damage. On the other hand, exposure provides a rough estimate of the direct economic value exposed to potential impacts of climate change and is therefore also a basis for the assessment of loss and damage. Loss and damage is an expression that refers to damage costs and losses of economic assets or stocks, as well as consequential indirect effects on economic flows, such as on GDP or consumption.

Consistently to the definition provided for RR, residual damage costs (RDCs) are the impact and damage costs after all desirable and practical adaptation actions have been implemented.

To sum up, financial strategies and instruments for adaptation and resilience can be categorized as in Table 2.1.

Table 2.1 Noncomprehensive list of financial strategies to address the variables of the risk and vulnerability functions.

Type of financial strategy/ instrument	Dimension addressed	Effect on the affected dimension	Effect on vulnerability/ risk
Disincentivizing investment in areas with high risk/hazard factors/scenarios/CC impacts	Exposure (e)	f' (e) < 0	ΔV < 0
Reducing sensitivity to CC impacts	Sensitivity (s)	n/a	n/a
Investment in adaptive capacity to CC (depends on residual adaptive capacity)	Adaptive capacity (ac)	f' (ac) > 0	ΔV < 0
Long-term capital budgeting	All	n/a	ΔV < 0
Risk transfer through insurance instruments and adaptation or catastrophe bonds	RR	Min RR	ΔR < 0

RR, residual risk; *CC*, climate change.

2.4 Sustainable finance and adaptation targets: in search of consistency

Most efforts for integrating climate aspects into financial decision-making refer to low-carbon transition, including carbon pricing and accelerating other mitigation activities. Only a limited attention has been paid so far to expected physical impacts from climate change and their relationship to the financial system, institutions, and economy (GCA Global Commission for Adaptation & UNEP, 2019), with some exceptions.

Nevertheless, a financial system is expected to convey functions that can benefit from the availability and quality of information on climate change physical impacts, vulnerability, risk, and potential damage. Levine (2005) mentions among others some functions that show a significant consistency with the concepts discussed earlier (in brackets in the list that follows):

1. producing information ex ante about possible investments and allocating capital *(vulnerability assessment)*;
2. monitoring investments and exerting corporate governance after providing finance *(monitoring the effectiveness or efficiency of adaptation).*

Additionally, a financial system can also convey functions that can support and benefit the effective implementation of climate change adaptation and DRR policies:

1. facilitating the trading, diversification, and management of risk (*risk transfer and management*);
2. easing the exchange of goods and services (*recovery after the occurrence of natural disaster/catastrophe*).

All these functions show a significant consistency with or dependency from the targets of climate adaptation and vulnerability or risk mitigation public policies. Actually, in delivering some of its distinctive functions, the financial sector could provide tools for addressing some pivotal aspects of adaptation and DRR policies.

The concept of sustainable finance has evolved as part of the broader notion of business sustainability over the last decades (Whiteman et al., 2013). As it happened in the domain of corporate social responsibility and nonfinancial reporting, the evolution in finance has seen the classical target of creating shareholder value evolving to the concepts of stakeholder value or the triple bottom line (Senge, 2008; Slaper & Hall, 2011) where nonfinancial metrics become a requirement to assess the performance of an organization. Reflection on sustainable finance has been "gradually shifting from short-term profit towards long-term value creation" by broadening the time horizon from short to long term (Schoenmaker, 2017). The most advanced stage of sustainable finance takes some distance from the approach of risk minimization, looks at the whole community as a target, and posits the creation of common good or shared value (Porter et al., 2011; Tirole, 2017), usually meaning that the common good is shared by and beneficial for all or most members of the community.

A typology for sustainable finance (Schoenmaker, 2017) considers four aspects in the assessment of the general orientation of a financial system:

1. value created,
2. ranking of the three factors determining value creation (i.e., financial, environmental, and social values),
3. optimization method, and
4. time horizon.

The three resulting levels of sustainable finance show an increasing emphasis on the participation of environmental and social components to the creation of total value depending on the minimum threshold set for nonfinancial components and a progressive widening of the time horizon.

Even though there is no explicit room to incorporate vulnerability and risk dimensions in the classification of sustainable finance (Schoenmaker & Schramade, 2019), we know that aiming to orient the investment decisions by financial institutions, the benefits of a potential risk reduction or increase in adaptive capacity should be internalized and turned to value.

A few enabling conditions can be set to favor the increase in AFI and the total amount of investment in targeted activities related to building resilience, based on the literature review:

1. Vulnerability and risk assessment outcomes rely on a time horizon consistent with the investment life span.
2. Vulnerability and risk assessment provide incentives to invest based on the consideration of the values of the variables of ac, s, e, as discussed previously.
3. The investment target belongs to vulnerable sectors or geographical regions affected by climate impacts (EBRD, 2018; EU TEGSF, 2020b).
4. The target and investment life span and its returns on investment are consistent with the vulnerability and risk assessment timeframe and their scenarios (EBRD, 2018).
5. The investee companies provide a sound set of climate-related financial disclosures in the field of adaptation reported on a yearly basis (TFCFD Task Force on Climate-related Financial Disclosures, 2017; ASAP, 2020).
6. The preference of groups of investors (private and institutional ones) on duration and expected ROI includes social and personal responsibilities and values (Riedl & Smeets, 2017), for instance, in the case of socially responsible investment funds (Omran & Ramdhony, 2015; Hong et al., 2016).

The benefits deriving to a targeted investment from actions achieving risk or vulnerability mitigation, or increased adaptive capacity can be either private or public/shared.

Private benefits accrue to the specific investment target or asset addressed by reducing risks associated with it, avoiding losses and directly or indirectly contributing to reducing costs and increasing related profits. Under this point of view, theoretically, no particularly advanced form of sustainable finance is strictly needed to deliver investment decisions being consistent with the prevalent targets of any adaptation or risk mitigation policy.

Public/shared benefits are relevant mainly under two points of view: they contribute to achieve public policy targets also by means of positive externalities or *spillover* effects; and they meet the criteria that financial institutions individually or in coalition have set aiming at delivering in bundle with financial returns also social, environmental or shared values (SEV).

In both the cases with private and public benefits, two essential conditions for all financial decisions have to be met:

1. returns on investments in vulnerability or risk reduction should be at least in line with prevailing market returns (that consider appropriately enough the associated risks and estimate their costs); and
2. the timespan of investment and the actual expectations of investors to get returns in the short run that justify the investment—which contribute to the decision whether or not investing in risk or vulnerability mitigation.

By looking at the three simple distinctive optimization functions for each of the three levels of sustainable finance proposed by Schoenmaker and Schramade (2019), we suggest how our conclusions on vulnerability and risk can be considered. Since the Schoenmaker and Schramade (2019) approach uses different conceptual categories in addressing the problem of risk management and climate change than the ones we discussed when introducing the concepts of risk and vulnerability, we provided qualitative comments on the objective functions.

The authors suggest a framework, including three incremental stages of sustainable finance (typologies), showing the characteristics presented in Table 2.2.

Interventions on adaptive capacity and exposure can be modeled as a form of vulnerability and risk mitigation. We have shown that adaptive capacity and exposure are the main variables where investment can provide effects being able to reduce the risk and vulnerability associated with specific assets. Changes in the vulnerability of assets targeted by investment decisions of financial institutions can affect the objective functions presented for a sustainable financial framework.

Particularly, for the first stage (SF 1.0) a decrease in vulnerability can help reduce losses (including economic ones) and risks for the investment by influencing the profit maximization and risk minimization functions. The type of vulnerability addressed in this case is mainly referred to the specific target of the investment under investigation, since the weight of positive externalities, spillovers, and public goods

Table 2.2 The three stages of sustainable finance according to Schoenmaker and Schramade (2019).

SF 1.0—Profit maximization, while avoiding "sin" stocks	Where FV = financial value = expected current and discounted future profits, and SEV = social and environmental values.Here the principal goal of the investor is to max the financial value assuring at the same time a minimum level of social and environmental values	max $FV = F(\text{profits, risk})$ subject to $F'_{\text{profits}} > 0$, $F'_{\text{risk}} < 0$, $SEV \geq SEV^{\min}$
SF 2.0—Internalization of externalities to avoid risk	p stands for the privately discounted value of the social and environmental impacts.Here the principal goal of the investor is to max the integrated value (IV), including at the same level the financial, environmental, and social components of value. The SEV shall not decrease over time. The success of the investment, however, might theoretically depend only on a good financial performance that outpaces the social and environmental performance of the investment	$IV = FV + SV^{p} + EV^{p}$max $FV = F$ (profits, integrated risk) subject to $F'_{\text{integrated profits}} > 0$, $F'_{\text{integrated risk}} < 0$, $SEV_{p}^{t+1} \geq SEV_{p}^{t}$

(Continued)

Table 2.2 (Continued)

SF 3.0—Contributing to sustainable development, while observing financial viability	Here the priorities of the investor change, even though both the financial and nonfinancial components of the investment keep their relevance in the final outcome. The principal goal of the investor is to max the SEV: thus no compensation is possible between the financial performance and the SEV performance of the investment. Additionally, the selection of a fair financial return makes particularly important to look after an effective financial performance that can be modeled as in SF 1.0 as far as profit max and risk min are concerned	max SEV = f (impact, risk) subject to $F'_{\text{impact}} > 0$, $F'_{\text{risk}} < 0$, $FV_{t+1} \geq FV_{t+1}^{\min}$ The financial viability or minimum financial value can be presented as: $FV_{t+1}^{\min} = (1 + r^{\text{fair}})\, FV_{t+1}^{\min}$ where $r^{\text{fair}} \geq 0$ is a fair financial return for one period

Source: Author elaboration on Schoenmaker, D., Schramade, W. (2019). Principles of sustainable finance. Oxford University Press.

delivered by risk mitigation and climate change adaptation measures are negligible since they would only participate in ensuring a minimum quantity of SEV.

For the second stage (SF 2.0), it is worth focusing on the integrated approach that calls for the consideration of integrated profits and risk, deriving from the combined financial, environmental, and social performance of the target of the investment. Therefore a reduction in vulnerability or risk could result in a reduction of more than one category of risk (e.g., social and environmental risks). The preferred measures will be the most effective in reducing the integrated risk value: either cobeneficial solutions or measures particularly effective on one component of integrated risk will be preferred. An intrinsic issue in the practice (if a minimum threshold for each type of risk is not set) is that some risk categories are left unaddressed because the ROI on their reduction is too low and unattractive for private investors. The management or transfer of those risk categories would be up to public policy and expenditure, or blended finance (Levenson Kehoane & Madsbjerg, 2016). The condition of SEV increasing over time allows factoring in also the benefits accruing from positive externalities and spillovers.

For the third stage (SF 3.0) the objective function to be maximized directly refers to the SEV generated by an investment that becomes *per se* the ideal target for the investor. In theory, here all risk mitigation and climate change adaptation measures addressing risk and vulnerability that show systemic effects significantly help to meet the maximization objective of the investor. The creation of shared value directly contributes to the primary maximization objective of the function. At the same time, the condition of increasing financial value (FV) over time, due to the presence of a "fair" (thus relatively low) return rate, makes all the measures aimed at increasing the resilience of the target of the investment under the private point of view even more important than under SF 1.0, since return rate can no longer play a significant role in determining the success of the investment itself. As a consequence, here the potential of increasing the FV by means of reducing expected losses from climate change and minimizing the risk for the target associated with climate change becomes of primary importance for ensuring a satisfactory performance of the investment. In line with the suggested form of the objective function, it is assumed that total value is maximized when:

$\max V = (1-\alpha)F + \alpha T = F + \alpha(S+E)$where V stands for the overall value maximized by the firm, F for financial value, T for total value ($T = F + S + E$), S for social value, and E for environmental value.

The evolution of SF can be interpreted as a change in the relative weights α and $(1-\alpha)$ of the financial and total (social and environmental) values in the consideration of investors. The level of social and environmental values is expected to increase as the financial system evolves toward SF 3.0 (Schoenmacher, 2016).

2.5 Preliminary conclusions and open issues

Finance has been historically pricing the risk of future cash flows for valuation purposes. The financial sector started considering the risk of adverse events originating in those companies where investments are concentrated since the early stages of the evolution of sustainable finance recalled previously.

However, in times of uncertainty about environmental risks and the relative policies, the practice of risk management can help to deal with uncertainties if appropriately adjusted and applied on a coherent timeframe. Scenario analysis is commonly used in climate science, but it has also become increasingly used to assess risk and provide valuation of investments under different climate scenarios (Caldecott et al., 2014). However, until now this mainly happens in the field of carbon pricing.

Finance deals with anticipating decisions and situations expected to occur in the future and incorporating public expectations into the valuations performed today for investment decisions, the returns of which materialize in the future.

Typically, finance reacts to policy announcements when addressing long-term investment decisions. However, it is hardly the case in our field of investigation. A distinctive problem with the financial system in addressing effectively environmental and particularly climate change challenges that have been largely investigated in literature is the one of "short-termism".

The time horizon to be considered in investments embedding measures for climate change adaptation needs to be long enough to allow for climate policies to be effective. This requires a profound change in the way in which financial results are assessed and managed that has been famously called for by major asset managers (Fink, 2016).

We mentioned that aligning the time horizons of investments with the ones of vulnerability and risk assessment in the case of climate change can be seen as an enabling condition for mobilizing profitable investment showing more consistency with adaptation policy targets.

Additionally, the huge debate on discount rate for climate change investments (largely developed concerning carbon pricing) deserves to be considered aiming to set a value of this rate for the long run that typically is lower than the prevailing discount rate offered by financial institutions, usually set in between 3% and 5% (Goulder & Williams, 2012; Weisbach & Sunstein, 2018; Giglio et al., 2018).

We did not enter the huge debate on metrics for adaptation here, but we suggested that risk and vulnerability assessment can convey information compatible with the needs of financial decision makers when evaluating the worthiness of an allocation of finance in a targeted initiative. More efforts on developing standardized metrics and indicators may help in supporting the spreading and application in daily work. Often, inconsistencies are found between public policies and the vulnerability and risk minimization objective mentioned that cause investment decisions adverse to adaptation and DRR targets and could require to be adjusted.

Literature proposed to use incentives and focused actions aimed to shift investment from short-term allocations to longer term ones, especially in the most vulnerable sectors. Recently, some categories of economic sectors, industries, and world regions have been framed, based on their exposure to climate change risks or on their vulnerability, or use of climate services (IPCC, 2014; UNFCCC, 2016; TFCFD Task Force on Climate-related Financial Disclosures, 2017; EU TEGSF, 2020a; Climate Bonds Initiative, 2019; EU-MACS, 2017; MARCO, 2018).

Some initial alignment between actions aimed to deliver more resilient assets to climate change and consistent or supportive financial decisions is now possible after studies concerning the vulnerability to climate change impacts and the associated risk in different economic sectors, industries, and world regions have been issued.

Delivering an alignment remains, however, a challenge that requires adjustments and improved dialog among scholars and professionals involved in the different domains at play.

Assuming that a very advanced approach to sustainable finance is adopted by significant groups of institutional investors, public organizations, and coalitions committed to sustainable finance, the delivery of good adaptation results, or significant reductions in risk or vulnerability could meet the goal to maximize the SEV under the condition to ensure a competitive level of FV and may deliver public positive externalities, the desirable effects of which could justify the implementation of policies and schemes aimed to shift finance to proresilience investment.

References

Archer, M. (2019). *Sustainable finance. Oxford Bibliographies in Environmental Science*. New York: Oxford University Press. Available from https://www.oxfordbibliographies.com/view/document/obo-9780199363445/obo-9780199363445-0117.xml?rskey = PztHBT&result = 1&q = sustainable + finance#firstMatch.

ASAP (Adaptation SME Accelerator Project), (2020). Adaptation Solutions Taxonomy. https://lightsmithgp.com/wp-content/uploads/2020/09/asap-adaptation-solutions-taxonomy_july-28-2020_final.pdf.

Bourdieu, P. (1986). The forms of capital. In J. Richardson (Ed.), *Handbook of Theory and Research for the Sociology of Education* (pp. 241–258). Westport, CT: Greenwood.

Caldecott, B. L., Tilbury, J., & Carey, C. (2014). *Stranded assets and scenarios. Discussion paper. Smith School of Enterprise and the Environment.* Oxford: University of Oxford.

Climate Bonds Initiative, (2019). Climate resilience principles a framework for assessing climate resilience investments. *Climate Bonds Initiative*. September 2019

Dufourmont, J., Papù Carrone, N., & Laxmi, H. (2020). Resilience and the circular economy: Opportunities and risks. Circle Economy.

EBRD, GCECA (2018). *Advancing TCFD guidance on physical climate risks and opportunities*. European Bank for Reconstruction and Development. European Program Climate ADAPT. Available from http://427mt.com/wp-content/uploads/2018/05/EBRD-GCECA_final_report.pdf.

EC (European Commission), (2018). Communication from the Commission to the European Parliament, the European Council, the Council, the European Central Bank, the European Economic and Social Committee and the Committee of the Regions Action Plan: Financing Sustainable Growth. COM/2018/097. Available from https://eur-lex.europa.eu/legal-content/EN/TXT/?uri = CELEX:52018DC0097.

EU TEGSF (Technical Expert Group on Sustainable Finance), (2020a). Taxonomy: Final report of the Technical Expert Group on Sustainable Finance.

EU TEGSF (Technical Expert Group on Sustainable Finance), (2020b). Taxonomy Report: Technical Annex.

EU-MACS, (2017). (European Market for Climate Services). Review and analysis of climate service market conditions. H2020 – SC5-03b Research and Innovation Action. Grant agreement 730500. Available from http://eu-macs.eu/wp-content/uploads/2017/07/EU-MACS-D11_CLIMATE-SERVICE-MARKET-CONDITIONS.pdf.

Fink, L. (2016). A Fundamental Reshaping of Finance. Blackrock letter to CEOs. https://www.blackrock.com/corporate/investor-relations/larry-fink-ceo-letter.

Füssel, H., & Klein, R. J. T. (2006). Climate change vulnerability assessments: An evolution of conceptual thinking. *Climatic Change*, *75*, 301–329. Available from https://doi.org/10.1007/s10584-006-0329-3.

GCA (Global Commission for Adaptation) & UNEP, (2019). *Driving Finance Today for the Climate Resilient Society of Tomorrow: For the Global Commission on Adaptation*. United Nations Environment Programme, July 2019.

Giglio, S., Maggiori, M., Rao, K., Stroebel, J., & Weber, A. (2018). Climate Change and Long-Run Discount Rates: Evidence from Real Estate. Chicago Booth Research Paper No. 17–22, Available from SSRN: https://ssrn.com/abstract = 2639748 or https://doi.org/10.2139/ssrn.2639748.

Goodwin, N. (2003). Five Kinds of Capital: Useful Concepts for Sustainable Development. Working Paper No. 03–07. Tufts University

Goulder, L. H., & Williams, R. C. (2012). The Choice of Discount Rate for Climate Change Policy Evaluation. Resources for the Future Discussion Paper No. 12–43. Available from SSRN: https://ssrn.com/abstract = 2145578 or https://doi.org/10.2139/ssrn.2145578.

Hong, L., Li, F., & Xu, J. (2016). Climate risks and market efficiency. *NBER Working Paper 22890*.

IPCC, (2014). Climate Change 2014: Impacts, Adaptation, and Vulnerability. Part A: Global and Sectoral Aspects. Contribution of Working Group II to the Fifth Assessment Report of the Intergovernmental Panel on Climate Change [Field, C. B., V. R. Barros, D. J. Dokken, K. J. Mach, M. D. Mastrandrea, T. E. Bilir, M. Chatterjee, K. L. Ebi, Y. O. Estrada, R. C. Genova, B. Girma, E. S. Kissel, A. N.

Levy, S. MacCracken, P. R. Mastrandrea, and L. L.White (eds.)]. Cambridge University Press, Cambridge, United Kingdom and New York, NY, USA, pp. 1132.

Levenson K. G., & Madsbjerg, S. (2016). The innovative finance revolution. Private capital for the public good. Foreign Affairs, July/August 2016.

Levine, R. (2005). Finance and growth: Theory, mechanisms and evidence. In P. Aghion, & S. N. Durlauf (Eds.), *Handbook of Economic Growth*. Elsevier.

MARCO (Market Research for a Climate Services Observatory), (2018). Final definition, taxonomy and report. Research and Innovation Action (RIA). Grant agreement No 730272. Available from http://marco-h2020.eu/wp-content/uploads/2020/01/MARCO_D2_3_Final_Definition_Taxonomy_and_Report.pdf.

Meadows, D. H. (1999). *Leverage points: Places to intervene in a system*. Hartland, VT: The Sustainability Institute.

Omran, M. A., & Ramdhony, D. (2015). Theoretical perspectives on corporate social responsibility disclosure: A critical review. *International Journal of Accounting and Financial Reporting*, *5*, 38–55.

Porter, M., & Kramer, R. (2011). Creating shared value. *Harvard Business Review*, *89*(1), 2–17, 2011.

Schoenmaker, D. (2017). Investing for the common good: A sustainable finance framework. Bruegel Essay and Lecture Series.

Riedl, A., & Smeets, P. (2017). Why do investors hold socially responsible mutual funds? *Journal of Finance*, *72*(6), 2505–2550.

Schoenmaker, D., & Schramade, W. (2019). *Principles of sustainable finance*. Oxford University Press.

Senge, P. M. (2008). *The necessary revolution. How individuals and organizations are working together to create a sustainable world* (p. 2008) New York: Doubleday.

Slaper, T. & Hall, T. (2011). The triple bottom line: What is it and how does it work?. *Indiana Business Review* [online]. Available from <http://www.ibrc.indiana.edu/ibr/2011/spring/article2.html> Accessed 23.08.2020.

TFCFD (Task Force on Climate-related Financial Disclosures), (2017). Recommendations of the Task Force on Climate-related Financial Disclosures.

Tirole, J. (2017). *Economics for the Common Good*. Princeton: Princeton University Press.

UNFCCC, (2016). Aggregate effect of the intended nationally determined contributions: An update. *Conference of the Parties. Twenty-second session*. Marrakech, 7–18 November 2016. Item X of the provisional agenda.

Weisbach, D., & Sunstein, C. R. (2018). Climate change and discounting the future: A guide for the perplexed. Reg-Markets Center Working Paper No. 08–19, Harvard Public Law Working Paper No. 0820, Harvard Law School Program on Risk Regulation Research Paper No. 08–12, Available from SSRN: https://ssrn.com/abstract = 1223448 or https://doi.org/10.2139/ssrn.1223448.

Whiteman, G., Walker, B., & Perego, P. (2013). Planetary boundaries: Ecological foundations for corporate sustainability. *Journal of Management Studies*, *50*(2), 307–336.

CHAPTER 3

Resilience stabilization: the role of economic issues in resilience practices

Giulia Pesaro
REsilienceLAB/Co.O.Pe.Ra.Te. Ldt, Italy

3.1 Toward an operational concept of economic resilience for territorial and community development

As already mentioned in the previous chapters, there are many possible definitions of resilience. In general terms the meaning that we want to emphasize here is that of the ability of a subject, an element, or a system "to resist pressure and distortions and be able to return to equilibrium." An equilibrium is not necessarily the initial one but might be a new one, determined by the adjustment process triggered by the need to reduce the pressures produced during the crisis and react. In a proactive perspective, during the adjustment process the territorial systems and the related communities can learn more about what is available and what is needed in terms of resources and capabilities. Such learning process is the basis for the development of a new possible equilibrium, result of an increased awareness of acquired or reinforced competencies, knowledge, and strengths to deal with ever-changing reference contexts, functional to increase the capability to cope with new events and changes possibly anticipating future crisis.

Change and transformation, together with rebalance and learning, describe a dynamic framework in which the discussion about the economic resilience and the contribution of economic approaches and tools to territorial resilience needs to be discussed and better understood. This is a territorial and community resilience perspective, where environmental, social, and cultural heritage and assets are key resources for a transition also addressing sustainable development goals.

It is important to underline that in this chapter a general discussion about the economic approach to territorial and community resilience

[ECO]systems of Resilience Practices
DOI: https://doi.org/10.1016/B978-0-12-819198-9.00010-7

development and enhancement is offered, while specific examples and interpretation of operational approaches and tools will be offered in the following chapters, with particular reference to Chapters 8 and 14.

As far as territorial systems and their balances are main focuses and material references for discussing resilience, it is worthwhile to offer a first operational definition.

In terms of extreme synthesis (Pesaro, 2012), the reference is to a territorial system "understood" as a heritage, a complex system of resources that support the production and consumption models that characterize its qualities, strengths, and weaknesses. Parts of the resources are tangible and part intangible; some are renewable and some are nonrenewable or might require very long time periods to be regenerated; some are private and some are public or common goods.

Given that the use of available resources determines a system of costs (direct and indirect, private and social), the set of resources should in any case be used in an economic perspective, continuing functioning with less resources (Pratt, 2015). That is, "in the best possible way, in the smallest amount possible, to obtain the best possible result," therefore to maximize the flows of value—incomes, rents, and benefits—determined by their use in relation to costs while maintaining or increasing the capital (in a strictly economic reasoning).

Assuming this position also means satisfying the set of needs and demand for goods and services expressed by local communities by giving greater importance to conservation objectives and trying to maximize the value flows determined by sustainable long-term uses, which actually are bases for the transition to more resilient territories and communities, suitable to also develop new and effective sustainable development paths. On the other hand, this enables the economic thinking, considering two key discussion elements (Pesaro, 2015). The first factor is the main conceptual framework of the economic approach understood in its broadest sense, that is, a disciplinary perspective which deals with the ways economic subjects behave in their territorial systems. Territorial systems are, at their turn, dimentioned and characterized by the presence of a certain endowment of resources, feeding the demand and offer for goods and services and the related production and consumption models. Systems, on the one hand, use their own resources to meet the needs expressed by their reference community and, on the other hand, are open to exchanges to provide themselves with what is not available locally and exporting what is required in other territories.

The second element is the fact that communities express a complex system of requests that varies over time, not only in quantitative but also qualitative terms, with dynamics that can lead to changes in the very perception of needs and of what is recognized as a set of reference values and component of well-being.

From this derives a possible first interpretation for an operational concept of economic resilience, which is "the ability of a territorial system to recognize and use its own endowment of resources to sustain its community of actors and develop in a sustainable way in the long term." This focuses on the strengthening of the resources system, maintaining its variety and quality, and protecting its value and availability over time.

In such a perspective the followings become crucial: (1) the ability to reduce the intensity of use of material and energy resources and reduce the consumption of nonrenewable resources in favor of renewable ones (at utilization rates compatible with natural regeneration capacities), (2) acting to reduce the impacts of climate change, (3) anticipate and better cope with the crisis which threaten its stability and capability to develop over time, (4) enhance community and intangible resources, (5) protect and enhance local well-being and quality of life in a sustainable development perspective. A set of goals presupposes a reorganization of its production and consumption activities and significant investments in research, development, and innovation as well as in the strengthening of human and social capital.

According to such an approach, working on resilience in an economic viewpoint requires two main capabilities. The first one is the capacity to adequately assess the territorial system of resources, verifying the related potentials and the possible scarcities based on the needs of local communities, and to produce or acquire what is needed to reduce weaknesses (also in terms of capacities and competencies). The second one is the need to adequately work on the system of demand for goods and services and on the requirements expressed by the community components while promoting new or renewed sustainability and resilience approaches.

Indeed, it is crucial to strengthen awareness on the values of the local endowment of resources and on the value flows which can be created in the medium and long term if the heritage is used but not consumed and if the production and consumption models will be able to better integrate the values produced by protecting and making the natural and built environments and the community members evolve at the same time. A path also embodies three main objectives: (1) strengthening the recognition of the fundamental values of common resources and of human, social, and cultural capital, alongside those more directly considered as economic; (2) the ever-increasing development of activities capable of producing flows of values from the conservation and enhancement of resources (Stanley, 2020) and not from their consumption (especially with reference to non-renewable material assets); and (3) to enhance the resources, the capacities, and the skills of communities to better cope with crisis and endogenous and exogenous shocks.

In the setting of objectives, actions, and investments, be these public or private, the protection and promotion of public and social resources and their quality are fundamental factors of what today should be identified as resilience capacity, or resilient development, of a territory. Following an approach that looks at the production of economic values as a measure unit for territorial development, practices enabling and enhancing territorial community resilience should be considered with equal dignity and importance of the more traditional measures of economic growth (like GDP, average incomes, etc.). An awareness implies a consequent adaptation, or rather innovation itself, of the decision-making processes linked to investments for development and the evaluation of their performance. A renewed economic perspective able to sustain, in the medium and long terms, a growing system of interconnected and more and more stable resilience practices, thus laying the foundations for the development of an overall economic thinking and approach accompanying the development of territorial resilience as a general action framework, like it already happens for sustainable development looking at green, sustainable, circular, and even bioeconomics.

Of course, it is important not to forget a different interpretation for economic resilience which originates from the definition of resilience as the opposite of vulnerability. As more vulnerable a system is, the more it is exposed and fragile to the negative impacts deriving from sudden changes, disasters, and states of crisis (Menoni, 1997). Should such events occur, the loss of resources and well-being will be directly correlated to the vulnerability of the system. A concept of loss, from an economic point of view, can also be observed in the displacement of productive activities, economic opportunities and, therefore, people in other territories considered more favorable. A dynamic has been clearly seen during different natural and man-made crises in the last 20 years.

Consequently, a system is all the more resilient the more it is equipped with the resources and capacities needed to positively deal with the negative impacts of crisis and shocks, reducing as much as possible the related damages and losses both in terms of quantity and time (i.e., the length of the period necessary to recover and get back to stability).

In this logic, economic resilience is the ability of a territorial system to react constructively and to adapt to endogenous and exogenous shocks and changes activating resources, means, skills, and solutions in the economic domain. This is such a way that, after a certain period of time, the system will be able to achieve again an equilibrium or, better, to achieve a new economic balance addressing economic exchanges and renovated relationships between demand and supply of resources, goods, and services.

If the actions to face crisis are developed in a positive and proactive way and possibly prepared in peace time, the final results might be a stronger and more prepared community, with increased competencies and capabilities to cope with further changes and shocks in a structural way, that is in the long term and within different (renewed) reference territorial frameworks.

New and more stable balances should also incorporate new goals referring to well-being, development, and value categories, which allow the system to sustain and, where possible, improve the standard of living over time. In positive terms the system is in some way stimulated to incorporate new categories of values and produce a renewal in the typology of needs and demand for goods and services, as the growth of the demand for environmental and social quality or the offer of continuous learning paths for all the territorial subjects. From here a virtuous circle is produced which, over time, also generates direct economic values in relation to the production of new activities, goods, and services functional to the new demand. The same endowment of resources that makes a territory resilient to negative changes can finally represent an important force factor to maximize the possible advantages deriving from events bearing positive effects, helping to make their effects structural in the medium and long terms (Aiginger, 2009).

3.2 Strengthening the role of economic thinking and tools in territorial resilience practices and the economic resilience of local communities

As mentioned previously, one of the difficulties which arise when dealing with economic resilience of territories and local communities is to identify a clear and operational system of economic-based tools without losing the richness of the whole of the micro, very specific and diverse examples that stem from the analysis of resilience practices and interventions. As it will be explained in more detail in the following chapters, where resilience practices in action will be analyzed and discussed, the success of the individual experiences depends very much on a system of factors and dynamics which characterize the crisis itself to be faced and the specific territorial situations.

The economic approaches and the related economic tools able to sustain and promote resilience practices and their achievement should be identified and selected according to (1) the specific vulnerabilities of the reference territorial system, (2) the available resources and capabilities to cope with the negative impacts of different crises, (3) the already available economic means and tools, (4) the capability of the active and target subjects to deal with economic instruments and to adequately consider economic sustainability as a crucial matter in the achievement of the expected results and, (5) the specific path chosen for enhancing local resilience.

In an ecosystem of practices perspective, where, as mentioned earlier, networking, sharing, and mutual learning are highlighted, the economic approach should be better considered. The economic issues and tools should be every time regarded as essential elements in the design of practices from the very first steps, because of the role of the economic feasibility and sustainability of the selected solutions. Moreover, the economic issues can be the target object of practices themselves, thus becoming the target intervention field. Finally, a system-based perspective could enhance the whole of the resilience actions and activities that take place in a certain territorial area. A system-level set of resilience goals, complemented by overall economic tools and solutions, can contribute to increasing the overall effectiveness and efficiency of the action.

These can help to give greater coherence to the set of sectoral interventions on the territory, both small and large in terms of investments, subjects involved, operational objectives, and expected benefits. The important thing is not the size but, precisely, the internal coherence and cooperation between local subjects in different practices and action fields, necessary and central to converge toward and contribute to achieving common goals.

In this logic, economic resilience appears to be a useful concept and a possible innovative interpretation for the evaluation of the dimensions and qualities of a territorial system. It also becomes a goal itself, leading local activities and investments toward new structural balances, more focused on quality and long-term sustainability objectives.

Following these reasoning a crucial node rises, which is how to understand and evaluate the different options and select the most suitable ones for each specific need in a certain territorial and community framework.

On the one hand, the possibility to adequately evaluate the weaknesses and the vulnerabilities characterizing an area could be the basis for creating the substratum for a more homogeneous and effective action system, as the system view could enable the identification of overall tools and solutions suitable for facing similar or complementary issues. In this way it is possible to better identify the elements and conditions to be produced, integrated, or strengthened to face the changes and make an active decision, hence orienting planning and investments in a strategic way at the local and even over local system levels.

On the other hand, the nature of the resilience-based projects and the associated expected results should be designed to make them more easily comparable and measurable both for their direct effectiveness with respect to a set of specific short-term objectives and for their coherence and contribution to the construction of system resilience in the medium and long terms. In this way a system of assessments can be conceived at the micro-level for evaluating the individual practice economic feasibility, implementability, and stability over time, and, at the same time, it is possible to better identify the overall territorial demand for economic means and tools to enhance or develop. The assessment reference framework should therefore be able to produce information and knowledge according to the following purposes:

- Produce knowledge frameworks that are able to provide a picture of the available territorial capital of resources (material and nonmaterial ones) in a long-term development perspective, looking at their characteristics, dimensions, dynamics, and role in terms of factors able to activate or produce collective and individual values. Such resource endowment should therefore be activated inside practices looking at the improvement of the quality of life and well-being at a community and system level and in a constantly evolving context, hence looking at territorial resilience enhancement.
- Identify and measure, if feasible, the vulnerabilities of the local system in relation to the different types of shocks and transformations to which it is exposed.
- Highlight the strengths and weaknesses of the system and the specific target fields or communities to highlight the coherence between the demand for resources and the local resources endowment. This is to

valorize the role and potentials of local resources and to enlighten, at the same time, the need for integrations or production of new key resources during the design of practices itself, enhancing their feasibility and implementability.

- Address decision-making processes and public and private choices, strengthening their mutual cooperation, to maximize the systemic effect of investments and the convergence of individual sectoral activities toward common and shared objectives.
- Highlight the quality of individual projects based on their ability to contribute to the achievement of the resilience and qualitative development objectives of the territory. This also considers a sustainable development perspective, minimizing the consumption of material and nonrenewable resources and producing a set of public and private values for the local community.
- Evaluate the effectiveness of investments on individual projects in relation to the system of expected direct results together with the contribution to systemic objectives.
- Highlight the relationship between public and private investments and benefits, in a short- and medium-long-term perspective.

It is here important to underline that in building and consolidating resilience through a system of practices at the local level, evaluation takes a strong strategic connotation, which strengthens its role. On the other hand, it needs attitudes and capabilities in terms of multidisciplinary approaches and the related variety of tools.

Economic variables, for instance, such as the variety and size of productive resources and activities or the availability of job opportunities, should be better considered according to what still happens in many programs and projects addressing territorial and community resilience (and will be described in Chapters 5 and 8). The potentials of the economic thinking for achieving resilience should be better acknowledged and valorized and not be relegated as a mere system of tools needed for implementing individual practices, many times not considering the possibility to activate practices at an over local level. In this discourse the perspective of territorial capital of resource, which is peculiar to the economic perspective, is particularly important to highlight the availability of productive resources which can be activated or might become themselves target of practices, as mentioned earlier. Economic tools might, for instance, sustain and enhance the economic role and the values of other territorial components, such as natural environments, landscapes, and cultural heritage (for instance, leading to new ways of looking to activities and practices in the field of agricultural production, tourism, and other sustainable resource fruition models).

Finally, on the one hand, rationality suggests that achieving local economic resilience goals, as well as better relay on the territorial capital of resources in a sustainable perspective, is clearly a win-win development strategy, on the other hand, it is still a challenge. This is because of the necessary reference to theories and practices which call for increasing capabilities of the designers of practices to work in an interdisciplinary perspective and better understand the synergies emerging from economic approaches and tools. Moreover, greater skills are required in evaluating and dealing with complex situations like those which characterize the economic interdependencies.

The analysis of practices, which follows in the next chapters, with examples from different approaches to the building of resilience in different action fields, involving different stakeholders and communities in different countries, demonstrates that this is still a challenge. An important challenge to be taken up which will be the reference for the discussions will follow in the final part of the book.

References

Aiginger, K. (2009). Strengthening the resilience of an economy. *Intereconomics*, 309–316, September/October.

Menoni, S. (1997). *Pianificazione e incertezza. Elementi per la valutazione e la gestione dei rischi territoriali*. Milano: FrancoAngeli.

Pesaro, G. (2012). *Analisi del territorio e sviluppo sostenibile*. Milano: Maggioli Editore.

Pesaro, G. (2015). Economie della resilienza per lo sviluppo dei sistemi locali. In A. Colucci, & P. Cottino (Eds.), *Resilienza tra Territorio e comunità. Approcci, strategie, temi e casi* (pp. 25–28). Milano: Fondazione Cariplo.

Pratt, A. (2015). Resilience, locality and the cultural economy. *City, Culture and Society, 6* (2015), 61–67.

Stanley, C. (2020). Living to spend another day: Exploring resilience as a new fourth goal of ecological economics. *Ecological Economics, 178*(2020). Available from https://doi.org/10.1016/j.ecolecon.2020.106805.

Further reading

Bristow, G. (2010). Resilient regions: Replacing regional competitiveness. *Cambridge Journal of Regions, Economy and Society, 3*(1), 153–167, March 2010.

Fankhauser, S., & Pearce, D. (2014). Financing for sustainable development. In G. Atkinson, S. Dietz, E. Neumayer, & M. Agarwala (Eds.), *Handbook of Sustainable Development* (2nd edition, pp. 446–459). Chapter 28, Cheltenham, UK and Northampton, MA, USA: Edward Elgar Publishing.

United Nations and Asian Development Bank. (2012). Green growth, resources and resilience. *Environmental Sustainability in Asia and the Pacific*, Report 01 January 2012.
Walker, J., & Melinda Cooper, M. (2011). Genealogies of resilience: From systems ecology to the political economy of crisis adaptation. *Security Dialogue, Special Collection: The Global Governance of Security and Finance*, *42*(2), 143–160.

PART II

Resilience practices observatory project

CHAPTER 4

Fostering resilience in local communities: the role of Fondazione Cariplo

Valeria Garibaldi and Elena Jachia
Fondazione Cariplo, Milan, Italy

Fondazione Cariplo is a leading Italian philanthropic foundation headquartered in Milan. It supports local communities and society as a whole in a number of ways, focusing its activities on four main areas, that is, the environment, arts and culture, scientific research, and social and human services, and fostering innovation across them. Established in 1991, during these last 30 years Fondazione Cariplo (FC) has funded over 30,000 projects of nonprofit organizations providing over 3 billion euros in grants:

- *promoting* sustainable development, the conservation of the natural environment and the cultural heritage, climate change adaptation actions and initiatives leveraging natural and cultural assets to drive growth;
- *supporting* scientific research and technology transfer and facilitating career development for young scientists;
- *backing* the development of solutions responding to social challenges such as population aging, welfare and well-being, social inclusion, improved health as well as better care of vulnerable individuals and social groups.

Resilience is defined as the capacity of a community to adapt to disturbance, surprise, and change in a way that still preserves its basic functions and ability to thrive (Magis, 2010). FC, consistently with its mission, starting from a specific call for proposal in 2014, has worked to build and enhance resilience within the communities it serves in its constituency (Lombardy and two provinces in Piedmont, Italy). It has done so to help them evolve and adjust to new environmental and social equilibriums in a sustainable way and preserve their identity as change comes about. To that end, it has invested resources to promote actions, interventions, and the development of *tools and methods*, while helping local players *build*

[ECO]systems of Resilience Practices
DOI: https://doi.org/10.1016/B978-0-12-819198-9.00013-2

knowledge and gain a deeper understanding of the issues at stake, being aware this is crucial to provide effective responses and solutions.

FC approach to resilience is grounded in the local system intended as the complex made up of the physical urban, and nonurban environment, of people/stakeholders and activities. As such its approach easily lends itself to be applied to a variety of different local contexts, players, and issues. *Building capacity* within the communities it serves to help them develop stable and enduring capabilities that enable them to cope with negative events or prolonged stresses has been another key area of focus for FC from the very beginning of its involvement in local resilience strategies/promotion. So, in addition to its financial support through call for proposals funding specific projects and initiatives, FC has also provided training and convening opportunities.

As the best way to promote resilience at the local level, FC has chosen to use calls for proposals, for a number of reasons. First, they can be focused on those issues that FC identifies as most relevant and needing urgent responses and resources. Second, when selecting projects submitted under a call, FC evaluates not only the goals and the directly expected outcomes of the proposed projects, but also their quality, their consistency with the call features, objectives, and target groups as well as the capabilities of the applicants/grant-seekers, and the method and work program proposed by them. Finally, since the first call for projects (Resilient communities) was issued in 2014, three more editions were held until 2017: this strategy allowed FC to address and support over time a variety of needs and actions but also stabilize overall results and boost the effectiveness of the actions taken. An approach which is actually considered quite unique in Italy.

These calls for projects were designed to build and enhance resilience in the communities FC serves. Communities have become increasingly vulnerable to environmental challenges and need to find ways to cope with them. However, resilience regards not just environmental but broader societal challenges that involve education, social and economic sustainability. Through its calls, FC has therefore solicited projects capable of strengthening a community's ability to react positively to change by putting in place solutions that catalyze local resources and skills and feature medium-term commitments as a way to ensure continuity as well as steadiness of outcomes.

In accordance with FC's mission and rules, only nonprofit organizations could apply for a grant. In addition, applicants were required to be engaged in the environmental field and in partnership with other local players. The project's goals had to be aimed at starting initiatives protecting the quality of local environment and its resources, advancing health and well-being of local communities, or promoting sustainable production models. The selected projects all spanned 2–3 years and were expected to generate lasting positive impacts.

The main features of the four *Resilient communities'* calls are summarized in Table 4.1.

Table 4.1 Goals and areas of focus of the "Resilient communities" call held from 2014 to 2017.

Call year	Goal	Focus	Examples of interventions
2014	Development of shared local initiatives aimed at *building and enhancing community resilience by preventing climate-related risks as well as other significant threats to the environment and sustainability*, such as unsustainable methods of production/consumption and land use	*Environmental management and conservation*	Prevention of risks and threats to the environment, protection of ecosystems, and reinstatement of environmentally protective functions
		Promotion of sustainable systems in the local economy	Promotion of new production and consumption practices, short food supply chains, local agriculture, local tourism, as well as local products, skills, and crafts

(*Continued*)

Table 4.1 (Continued)

Call year	Goal	Focus	Examples of interventions
2015, 2016, 2017	Development of initiatives *capable of boosting community resilience vis-à-vis vulnerabilities identified at the local level*	*Environmental management and conservation*	Prevention of risks and threats to the environment, protection of ecosystems, and reinstatement of environmentally protective functions, shared management of common goods and resources
		Quality of life, health, and well-being	Reduction and mitigation of pressure and pollution factors, promotion of food quality and safety as well as environmental quality of urban spaces
		Enhancement of local environmental resources and sustainable productions	Promotion of new production and consumption practices, short food supply chains, local agriculture, local tourism, as well as local products, skills, and crafts, sharing and circular economy

Source: Fondazione Cariplo internal assessment reports. Reference years: 2014–17.

Under the first two calls launched in 2014 and 2015, FC funded with 2.4 million euros a total of 26 projects out of 132 project proposals received. In 2016 and 2017 the funded projects were 28 out of 95 proposals received, granting 2.8 million euros. Over the 2016–20 periods 48 projects funded by FC got completed. These projects involved 124 project partners and 605 other local players. Table 4.2 shows key figures for the projects completed.

Given the size and scope of the strategic endeavor and financial resource, the need soon emerged for in-depth analysis and assessment of all funded projects, their overall results, and, more broadly, the effectiveness of the overall program and of the financial and organizational resources. So, FC funded an independent research team composed by Politecnico di Milano researchers (Osservatorio sulle pratiche di resilienza), the activities and insights of which will be presented in the next chapter.

Table 4.2 Key figures of projects funded and completed under FC "Resilient communities" calls.

	Reference year for the data collection		Grand total
	2016–19	2020	
Projects completed between 2016 and 2020	45	3	48
Total grants (in thousand euros)	4341	282	4623
Total worth of funded projects (in thousand euros)	7743	502	8245
Partners involved in the projects	116	8	124
Other local players involved in the projects	593	12	605
Mapped land (in ha)	531,768	28	31,795
Environmental vulnerabilities identified	398		398
Signatories to management agreements	2142	38	2180
Land covered by renewal/regeneration interventions (in ha)	401	60	460
Players involved in participatory planning	4724	2010	6734
Local short-supply chain producers involved in promotion events	786	5	791
People trained or involved in workshops	6087	668	6755
Total training days	1010	123	1133
Awareness raising events	775	77	852
Attendees at events	42,037	12,750	54,787
Schools involved in education activities	167	24	191
Participating students	12,118	2067	14185
Human resources engaged in the projects	294	30	324
People reached through project-related direct communication activities	474,899	13,497	488,396

Source: From Fondazione Cariplo internal assessment reports. Reference years: 2016–20.

Some features stand out as the main strengths of the funding program:

- The stability of the action over 7 years (4 calls and an average duration of projects between 2 and 3 years), which yielded a multiplicity of activities, initiatives, and experiences enriching and strengthening the local communities involved.
- The contribution to the establishment of a "community of practice" formed not only by the project partners, but also by the different stakeholders (associations, businesses, and local authorities) involved in the interventions, activities, and experiences over 7 years.
- The number and the variety of initiatives carried out within communities that were a fruit not only of the quality of the selected projects but also of the continuity of support that generated important synergies over time and among different communities.
- The learning processes which have been established among the members of the different communities and among different stakeholders as a result of the active involvement of project actors and of the dissemination and communication efforts by the project partners and FC itself.
- The results obtained through the stable medium-term funding strategy FC carried out, which has enabled a variety of partners/players to implement their actions under a coherent reference framework, but also FC itself to go through an evolutionary learning process—and these are key traits that distinguish the Foundation's approach in supporting the communities it serves.

Finally, the most prominent feature has been the involvement of the local population that took an interest in community resilience and was engaged in the promotion of sustainability and well-being, in the risk reduction practices, as well as the development of local sustainable production and consumption models. The degree of involvement of the local communities has actually been one of the main criteria used to judge and select the projects submitted under the calls. Building resilience means building the abilities of a community to positively react to risks and crises. A key way to do so is to focus common goods protection capitalize on them. This is what FC has done for the communities it serves, backing projects having that goal and providing a compass for communities in their search for new, stronger sustainable equilibriums.

Reference

Magis, Kristen (2010). Community resilience: An indicator of social sustainability. *Society and Natural Resources*, *23*(5), 401–416. Available from https://doi.org/10.1080/08941920903305674.

CHAPTER 5

Resilience practices observatory project: emerging phenomena and lessons earned

Angela Colucci and Giulia Pesaro
REsilienceLAB/Co.O.Pe.Ra.Te. Ldt, Italy

5.1 The Italian Resilience Practices Observatory

5.1.1 Resilience Practices Observatory: the project

In recent years, there has been a remarkable expansion in the use of the resilience concept and broader diffusion of resilience practices, making territorial and societal quality a central element in urban and territorial innovative actions and investments. Resilience has increasingly become a crucial feature of territorial and social systems and the communities of territorial actors, underlining the significance of active adaptive behavior (Burayidi et al., 2020; Roostaie et al., 2019).

The diffusion of resilience practices seems strongly correlated to the capability to make communities and territorial stakeholders aware of positive impacts deriving from a change promoted by administrative bodies or institutions and integrating a bottom-up perspective (Colucci, 2012; Galderisi & Colucci, 2018). Therefore the strength of resilience is in the growing multiplicity of concrete experiences and operational applications designed and implemented at the local and territorial levels on environmental urgencies (Berkes et al., 2000).

The Resilience Practices Observatory (RPO) arose from the intersection of two parallel paths: the REsilienceLAB association path aiming to promote and support the application of the resilience approach from theoretical to action intervention domain and the Resilient Communities program launched by Cariplo Foundation. RPO arose from the shared crucial urgency in understanding and evaluating the benefits deriving from transition/resilience practices geographies and highlighting promising instruments to improve the practices stabilization and upscaling (Colucci, 2018).

[ECO]systems of Resilience Practices
DOI: https://doi.org/10.1016/B978-0-12-819198-9.00009-0

Fondazione Cariplo (see Chapter 4), an important Bank Foundation that finances numerous environmental and social initiatives, launched a call for Resilient Communities. This call made local communities in the territorial area concerned by the funding activity of Fondazione Cariplo (the whole Lombardy region and a small portion of Piedmont) propose, imagine, and build up resilience practices. Fondazione Cariplo received a great number of project proposals for implementing resilience practices at community and territorial levels. On such basis, it has been proposed to build an Observatory with the goal of producing new knowledge founded on the collection, analysis, and interpretation of the activities, projects, and actions aimed at strengthening resilience of territories and communities in a long-term perspective. A place and a tool based on a interdisciplinary approach intended to become a discussion arena and a toolbox itself. The RPO officially started in April 2015. The project is supported and funded by Fondazione Cariplo. The lead partner is the Department of Architecture and Urban Studies (DASTU) of the Politecnico di Milano and the other partners are the Department of Science, Planning and Territorial Policies (DIST) of the Politecnico di Torino, the University Consortium for Socioeconomic and Environmental Research, and the association REsilienceLAB.

The second phase of Observatory (2017—19) modified the partnership: The lead partner remains the DASTU of the Politecnico di Milano and the other partners are the DIST of the Politecnico di Torino, the Department of Biosciences and Territory of the University of Molise, Lombardy Foundation for the Environment (Fondazione Lombardia per l'Ambiente), and the association REsilienceLAB.

5.1.2 Resilience Practices Observatory: the aims

Taking into consideration the set of action goals introduced previously, the working project of the RPO has been designed following three main elements. The first one relates to the emerging demand for integrating reflections about the complexity of the concept of resilience itself under a scientific and cultural profile with reflections on resilience practices, which represent the operational side. The second one is the need to integrate and deepen specific considerations about multiple interpretations and approaches to resilience in a robust interdisciplinary perspective to obtain a "new multidimensional understanding" of the resilience concept. Finally, the third one concerns the RPO's development as a network of multiple subjects that support, participate, observe, and implement the practices. The possible constitution of an inclusive and stable network, which takes together academia, research, professionals, communities, local authorities, and private and public institutions and bodies, proved to be an essential element for resilience practices support and exploration.

The aim of RPO is to give and produce conceptual frames supporting clear orientations in the complexity, highlighting possible aspects of conflicts and overlays. There is not a univocal "proposed definition of resilience practice" but a "concept of resilience practice," which was intentionally left as broad as possible, to contain a variety of initiatives that are based on different approaches and toolboxes.

A resilience practice is therefore conceived as the activation of a process. At the same time, the RPO assumes more rigorous and scientifically based definitions working on assessment methods and evaluation criteria. The focus is on processes and actions undertaken by the practices, which have to be able to strengthen the capabilities and the resilience properties of the system itself. The resilience of complex systems is therefore seen as a sum of capabilities of the system able to activate responses to phenomena of stress and shock, positively adapting to new conditions and maintaining its identity and its functionality over time.

The idea of an adaptation process, to better face and be prepared facing crises and shocks, has then become more and more popular and has been developed in projects and practices by communities, mainly in a bottom-up perspective. This is because:

- it implies a strategic and positive dimension. Communities and territories provide themselves with the features and skills needed to be able to overcome stress phenomena and getting, at the same time, new equilibrium and capabilities to take advantage of different scenarios;
- it is inclusive of multiple "uses" and declinations, in relation to contemporary phenomena, and be applied at many different territorial levels and sectors. Starting from the issue of territorial risks, both in terms of catastrophic events and in long-term perspectives related to health and well-being of populations, the concept finds more and more success in the local initiatives of communities. It is an effective approach and toolbox to deal with crisis phenomena related to economic, social, and governance changes;
- it involves a process approach and allows for the building of changing paths over time and in a strategic and long-term perspective. This by giving birth to development and learning paths and to the production of resources, skills, and expertise to anticipate future events;

- it implies a temporal approach that sees a fundamental element in the continuous development of resilience practices.

5.1.3 Resilience Practices Observatory: activity domains

The Observatory takes as its overall strategic objective the strengthening of the capacities of territorial subjects to enhance resilience (Fig. 5.1). This considers both all territorial entities, institutions, and communities involved in a long-term transition project, toward a stronger, aware and adaptive society, and the subjects themselves which animate the RPO. This is why the RPO project design has been based on the following main axes.

- *Mapping resilience initiatives at national level.* Here the concept of "mapping" refers to the understanding of characteristics that differentiate initiatives and to the understanding of the territorial areas mainly involved in the resilience practices. This proved to be an important step toward the better understanding of the potentials of the multidimensional nature of the resilience concept "in action," when applied to urban and territorial practices. The mapping required close integration between the construction of conceptual frameworks of reference and the direct involvement of the promoters as it was not possible to build up clear distinctions and to "position the practices in a map" without having a direct confrontation with the goals and expectations of the subjects directly involved in the project design or promoting it.

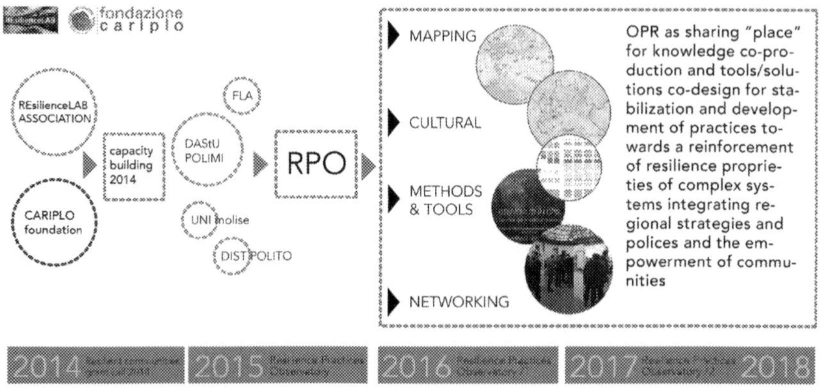

Figure 5.1 Resilience Practices Observatory: a synthetic timeline and the main four activity domains. *Credit: Angela Colucci.*

- *Develop a set of tools and design criteria to support the dissemination of resilience practices*. The goal is the building of new toolsets and project design criteria to support the expansion and reinforcement of resilience practices and the solutions design. These tools are therefore meant to be offered to actors that support the community planning and are identified as "promoters"; they can be economic and private subjects able to better understand the opportunities for innovation and economic local development; institutional actors, and the civil society players.
- *Produce developments in the cultural and methodological paths*. The focus here is on capacity building and innovation in knowledge approaches. The strategy relies on promoting scientific advancement about resilience and transition themes, based on applied research to territory and community. This is to produce advances in scientific research in terms of conceptual and methodological innovations.
- *Promoting the involvement of extensive actors' networks*. The definition of conceptual tools, both interpretative and for planning, will try to involve, in a process of mutual magnetization, the promoters and the community through the implementation of innovative and shared paths. This is based on a multisectoral and interdisciplinary perspective.
- *Enhancing the monitoring and the evaluation which are integrated into all work steps*. This with a particular attention to the capability to evaluate the overall resilience enhancement effects of the whole of the practices and looking at the goals of territorial policies.

The discourses on resilience concept and approaches, concerning the *cultural path*, have continually accompanied in a transversal and multisectoral way all the other activities of the observatory (Fig. 5.2). A path on the resilience concept reframing was then developed, to enhance the interdisciplinary approach and obtain a integrated and shared vocabulary to better discuss and exchange views with the Observatory stakeholders. Three main information and knowledge sources have then been considered:

- Scientific literature.
- Web and dissemination platforms and documents.
- Consolidated practices mostly related to risk prevention, climate change adaptation, environmental and sustainable planning, and urban design. This last activity field has proved to be the more challenging due to the variety of sectors/issues addressed and of the practices themselves.

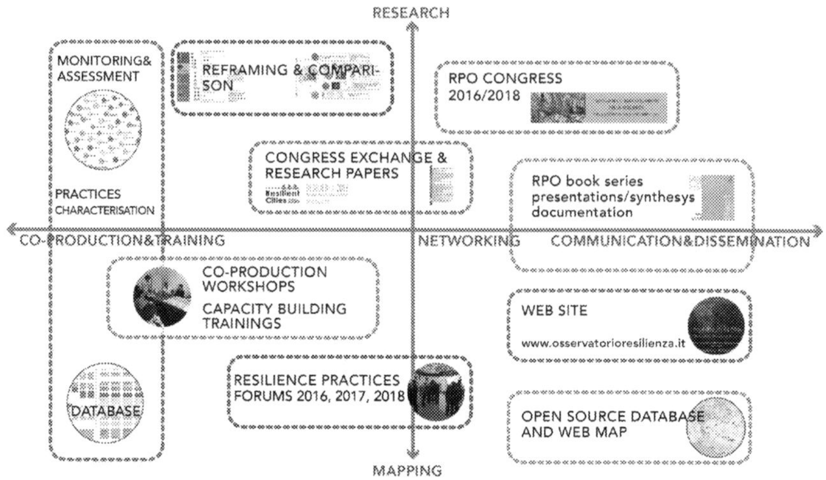

Figure 5.2 Resilience Practices Observatory: products and initiatives along the activity domains. *Credit: Angela Colucci.*

During such activity, called *the reframing phase*, internal debate and workshops were organized to explore all the opportunities and concepts deriving from engineering, ecosystem sciences, informatics (cybernetics), and psychological disciplinary approaches. A parallel activity was the collection and comparison between existing and consolidated practices related to main approaches to resilience (risks, climate change, adaptation, transition) and the connection between theoretical models and the panorama of practices.

RPO launched and organized several initiatives such as the Resilience Practices Forum oriented to the networking involving academia, research, professionals, communities, local authorities, and private and public institutions and bodies.

5.2 The Resilience Practices Forum and the knowledge coproduction activities

5.2.1 Resilience Practices Forums

After the starting of the RPO working activities, it has rapidly become clear the need to engage and dialogue with resilience practices, to better understand the widespread use of the "resilience" term. The practices already involved in the Fondazione Cariplo's calls for funding (mentioned above) were not considered sufficient to obtain a clear representation of the national situation. This is for two main reasons. The first one was the territorial limitation of the funding call, due to the institutional boundaries for the funding action as in the Fondazione Cariplo's statute, which considers only the Lombardy region and a province in Piedmont region. The second one was linked to the call itself, as the project design activities were addressed by the specific requirements needed to have access to the funding program. It was therefore essential to involve practices and experiences at national level, linked to different approaches to the resilience concept and action.

The RPO decided to organize the Resilience Practices Forum (RPF) initiative as a tool for the practices' engagement and, at the same time, a place for experiences exchange, for dialogues among project promoters, academia, and experts and institutional stakeholders (Figs. 5.3–5.5).

The RPF has then been configured as a path that connects different worlds. The first one is the world of the practices, or of the actors that promote and implement the practices which are mostly public and private bodies that support and promote the initiatives and processes. The second one is the world of research and innovation, represented by universities, practitioners, international networks, and national research centers.

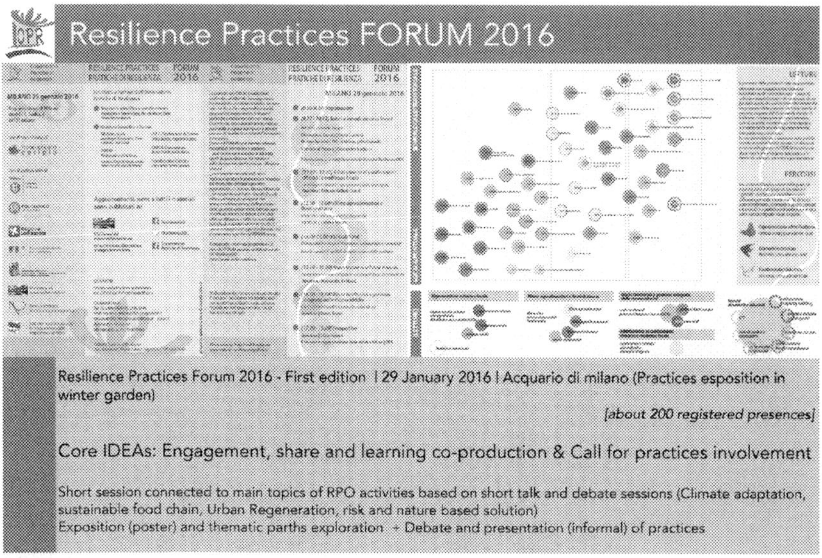

Figure 5.3 Resilience Practices Forum 2016 edition. *Credit: Angela Colucci.*

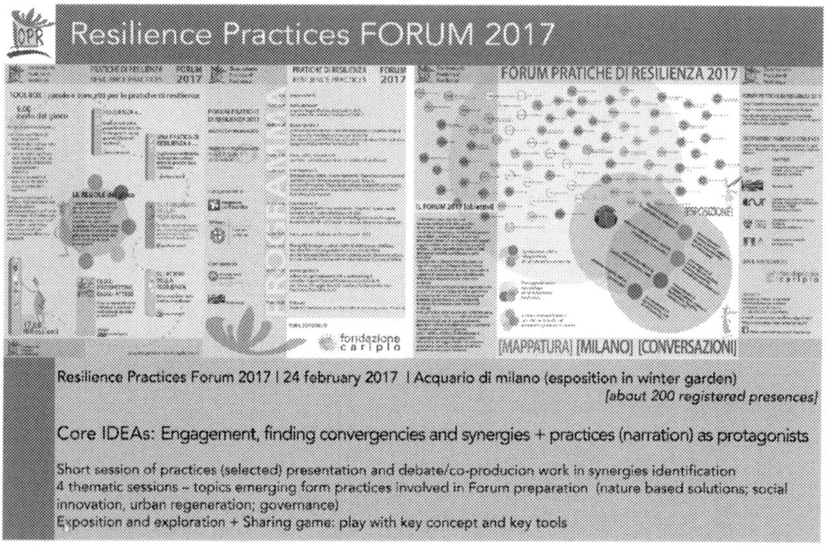

Figure 5.4 Resilience Practices Forum 2017 edition. *Credit: Angela Colucci.*

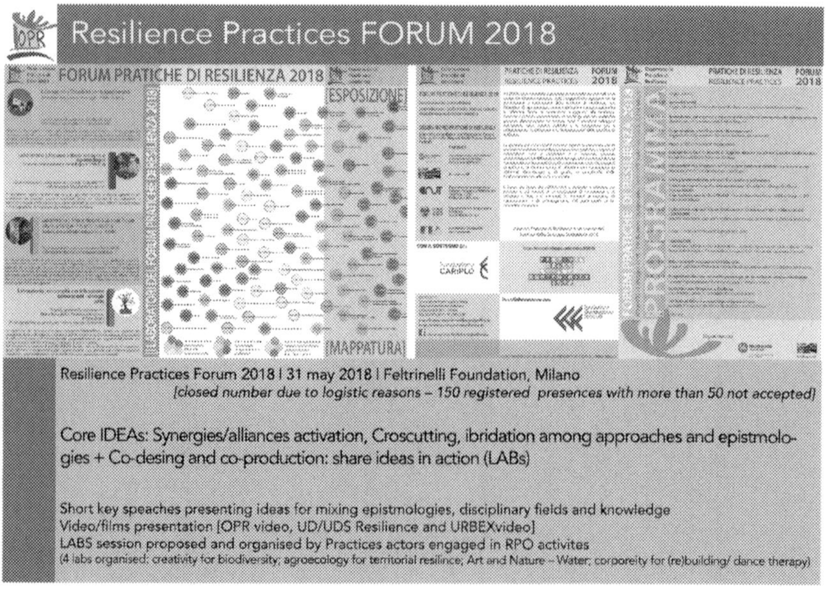

Figure 5.5 Resilience Practices Forum 2018 edition. *Credit: Angela Colucci.*

The goals of the RPF were many and ambitious, in the light of final goals looking at the enhancement, promotion, and support of the practices. The activities have therefore been organized to share and discuss knowledge and approaches, to enable and/or strengthen alliances and cooperation, to consolidate and innovate the working paths, and to develop criteria and methods for the RPO practices mapping.

The overall goals of the RPF were:

- to improve the knowledge and share experiences. The exhibition have been organized as moment of public presentation, comparison, and sharing between resilience practices;
- to solicit the debate. The RPF sessions promoted the debate in relation to the main issues and on resilience practices;
- to support alliances and cooperation. The RPF was meant to be a first step for the networking process involving and engaging active players in proposing, supporting, and developing initiatives and researches on and about the resilience practices;
- to develop instruments. From the RPF the RPO received crucial inputs for the development of principles, criteria, and methods for the interpretative frameworks, for innovative tools development to support knowledge, dissemination, and implementation of resilience practices (policies, guidelines, tools, and governance solutions);
- the dissemination and capacity building. The RPO promoted initiatives for the capacity building and dissemination of resilience practices.

Under such conditions the RPO launched three calls in 2016, 2017, and 2018 with a growing interest and participation of the resilience actors and practitioners at the local and national levels. The RPF has been conceived as a path able to connect different action spheres.

- the sphere of practices, or of the actors that implement the practices;
- the sphere of "supporting" actors, including private and public institutions and bodies that support and promote the initiatives and the resilience processes;
- the sphere of research and innovation, also by way of the involvement of international networks and national research centers;
- the sphere of future action, aiming at promoting the debate and the identification of further focus involving all the stakeholders of network.

The RPF, looking at it as a tool itself to enhance the resilience practices system, was characterized to be, mainly, the starting point of a process of networking and of exchange, enrichment, and shared reflection among a large range of actors and stakeholders involved, with different interests and roles in the promotion and implementation of resilience practices. The three RPFs have therefore been organized as processes of activities, first by communication and exchanges through digital devices and then in meetings in Milan (one for each call).

Three main phases have characterized the RPFs preparation and functioning:

- the organization phase, with the launch of a call for the collection of practices at the national level and the definition of the key topics for each event;
- the RPFs events held in Milan. The events hosted different moments of debate, comparison, and exchange between "practices" and posters exhibition. The debates, in particular, supported by some interventions from experts and practitioners, the practices themselves and discussions around emerging issues, led to the raising of the most common strengths and weaknesses of the resilience concept into action and helped to identify some main key issues and complexity aspects;
- the enrichment and dissemination of the results of the forums after the events themselves. In the following months, through different initiatives coordinated by the RPO, the debate on values, solutions, and tools proposed and implemented has always been carried on, to promote the visibility of the practices and maintain active the network, the participation, and the exchanges also after the meetings.

From the Forums, many inputs arise not only in terms of "networking" but also synergies activated among practices. About the feasibility of the projects or practices and the weaknesses and/or barriers to be overcome, the RPFs have been important occasions to have a confrontation with the practices designers and stakeholders. For instance, the RPFs activities permitted to light up the weaknesses noticed by the practice designers and by the implementing communities. The principal barriers emerged can be summarized as follows:

- institutional barriers and administrative/bureaucratic weaknesses producing timing problems and obstacles in decision-making and activities needed to make the practices be implemented;
- lack of financial means or of stability and continuity of the funding;
- scarce confidence in the feasibility and effectiveness of public–private cooperation needed to better face territorial criticisms;
- weaknesses in the capabilities in "building" governance processes (in particular in relation to "people" involvement and engagement) but also difficulties in the building of local and over local networks (difficult relationships with already existing practices and processes).

5.2.2 Resilience Practices Observatory Congress and coproduction activities

Rooted on the emerging barrier and weakness aspects, the RPO developed several activities for capacity building dedicated to codesign, community engagement, and practices stabilization. The capacity building and coproduction initiatives, launched from 2015 to 2019, could be summarized in main activities clusters:

- The workshops and coproduction seminars: for example, a series of workshops dedicated to the codesign and engagement tolls in which involved practices presented and discussed the "tools" used. These workshops represented a crucial instrument to improve the capacities of organizations because they shared the tools used pondering/discussing the success/barriers aspects found in engaging communities and stakeholders, shared more promising useful tools and increased their capacities in selecting the correct tools in relation to the action goals.
- The RPO website that includes an interactive web database (open-source data) could be explored according to some filters (e.g., size, criticalities, topics), specific guidelines and updates on founds call (and on economic sustainability), guidelines and link to international practices, guidelines on main topics addressed by practices. The RPO website provides an interactive environment and dynamic contents updated in relation to practice requests and needs (emerging from RPO activities).
- The RPO also promoted the production and diffusion of different media: for example, RPO launched during the RPF events produced videos on resilience or documents and materials of dissemination.

The "RPO Congress" also responded to the coproduction research approach (dynamic and by design, Fig. 5.6). In general, the two Congresses organized were addressed to open the discussion on resilience practices to international debate and engage multidisciplinary actors at the national level. The first Congress (in 2016) presented a more conventional framework engaging different experts approaching the urban and territorial resilience from different points of view. The second Congress (2018) aimed to innovate the academic congress format, including two "dialogues/talks" sessions to emphasize the role of actions and the exchange among researchers and practices with a knowledge coproduction approach. The Congress included two sessions of "conversations" engaging more than 30 practices from the national context presenting their experience and debating with researchers

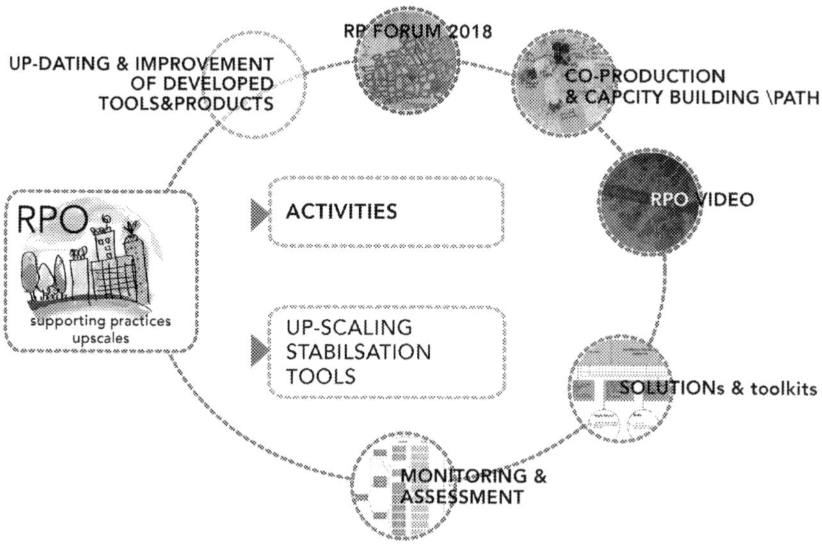

Figure 5.6 Resilience Practices Observatory: activities supporting the practices upscaling. *Credit: Angela Colucci.*

and experts on results, impacts, and barriers to improve the resilience capacities of complex territorial systems.

5.3 Exploring the resilience practices observatory practices

5.3.1 Resilience practices database

The collection of practices from both the Cariplo Foundation funding calls and the Resilience Practices Forums (RPF) networking calls formed a database of practices, aiming at enabling the creation of new networks among people sharing similar visions; finding/exploring new (interesting and innovative) practices; monitoring trajectories and trends from the field.

5.3.1.1 Defining resilience practices and databases

The RPO does not assume one specific definition of resilience but prefers to recognize many different definitions and approaches to resilience to include the more comprehensive possible range of practices (Fig. 5.7). The RPO assumes that the resilience of complex systems is the sum of capabilities of the system, enabling the activation of responses to face to stress/shock phenomena reaching, through processes of adaptation, new

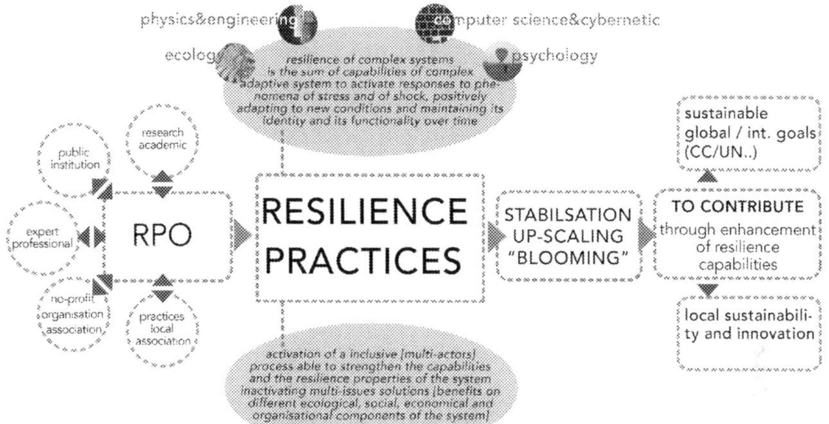

Figure 5.7 Role of resilience practices in the Resilience Practices Observatory project. *Credit: Angela Colucci.*

conditions maintaining its identity and its functionality over time. RPO developed a "concept of resilience practice," where process and actions have to strengthen the system's capabilities and the resilience properties of the system itself (Colucci, 2018).

The projects supported by Fondazione Cariplo (selected to be granted in the Resilient Communities Funding Calls in the years 2014, 2015, 2016, and 2017) constitute the RPO database setting's starting point. An extensive range of practices was engaged and included in the database during the codesign and coproduction activities (Congress, RPFs, workshops, and the other events organized by RPO): since 2015 up to December 2019, the RPO has engaged and explored more than 170 practices.

The RPO elaborated two databases:

- The RPO internal database (the one discussed in this paragraph) that includes 147 practices based on an ongoing process of database stabilization and processing.
- The RPO public database published on the RPO website: all the data are accessible, consultable online using a web map or explored through categories, and downloaded (the database is an open-source resource). The RPO public database includes 72 practices, and it is a "selection" of all the practices engaged and explored. The selection is based on the completeness, availability, and possibility of sharing the information of the practices (all information must be open-source or public) and

on the relevance of the single practices in contributing to achieving the dissemination and capacity building purposes of the RPO website.

5.3.2 Resilience practices: sources and typologies of practices

The primary sources of Practices engagement (Table 5.1) are the activities of the RPO and the Resilient Communities Program promoted by Cariplo Foundation. The RPO research database results from a necessary simplification process.

In general, the database photographs the situation as of December 2019 and includes only practices on that date that were verified and for which there was the necessary information to complete the many survey factors here presented.

Several practices (and related actors) participated in more than one RPFs or presented single initiatives connected to the principal "process" in different RPO events. For example, several practices launched in occasion of the Resilient Community program of the Cariplo Foundation have presented in the RPFs specific initiative or the results achieved in a specific phase of the process evolution. Therefore the presence of the Cariplo-funded practices could have slightly biased some analysis factors,

Table 5.1 Resilience Practices Observatory research database: resilience practices sources.

	Number	%	Number (total)	%
FC CR call 2014	12	8.16		
FC CR call 2015	14	9.52		
FC CR call 2015–16	14	9.52		
FC CR call 2015–17	14	9.52		
Tot. FC CR call			**54**	**37%**
RPO RP Forum 2016	41	27.89		
Forum 2017	16	10.88		
Forum 2018	19	12.93		
Total RP Forum			**76**	**52%**
RPO 2017	7	4.76		
RPO 2018	10	6.80		
Total RPO			**17**	**11%**
Total	**147**	**100.00**	**147**	**100%**

FC call is for projects funded by Cariplo Foundation in the Resilient Communities program; RPO RP Forum is for the resilience practices involved during the RP Forum events; RPO is for practices involved/included during other RPO activities.
The table shows the number of practices involved in relation to each year of Call/Event editions and in bold the total of practices in relation to the three main involvement sources.
Credit: RPO database, elaboration by Angela Colucci.

like the geographic location or some operational characteristics, because of the call's requirements. The RPO database anyhow integrates experiences from all over Italy.

5.3.2.1 Resilience practices typologies

The first aspects analyzed while processing the database regarded the identification and attribution of resilience practices typologies. The necessity to attribute a typology is primarily an operational need to better address and manage the heterogenous set of practices. Specifically, the primary attribution of "typology" was not intended as interpretative action but as initial characterization/identification of the practice in the moment of involvement in the RPO. Although partially correlated with the source from which they came, the adhering practices were characterized by some "similar" characteristics ascribable to the practice intentionality and action approach. The process's centrality led to the definition and characterization of the practices (precisely concerning the relevance of process—implicit or explicit—that catalyzed the practices). Over time, some have activated a social innovation process, while others were very focused on a specific intervention (generally moved by specific urgency to which to answer).

The typologies identified (Table 5.2) are:

- process: the activation of a local innovation process is the core of these practices (it is not possible to identify one specific initiative, but multiple interventions/activities activated in a complex local process);
- project/proposal: the core of social activation is a project or a proposal development (not or not yet started/implemented);
- process/action: the practices are characterized by a process in which the main focus is the activation of intervention on/in local contexts;
- action/intervention: the focus of the practices is the achievement of one specific goal/intervention in the local context and (if) the process activate is oriented to a specific and single goal/urgency;
- process/research: the practices are applied research path promoting the activation of local innovation processes;
- research: the "practices" are research paths focused on the resilience of the complex systems, but these do not directly activate a social/local process.

The practices are mainly based on processes or on a mix of processes and specific local actions, which have to be developed in the territories and among the communities' targets of the practices. This was expected, as the attention is focused on concrete operational practices. Still, it is

Table 5.2 Resilience Practices Observatory research database: resilience practices typologies and distribution.

	Number	Percentage
Process	46	31.29
Process/action	62	42.18
Project/proposal	2	1.36
Research	11	7.48
Process/research	16	10.88
Action/intervention	10	6.80
Total	**147**	**100.00**

Bold value highlights the total line.
Credit: RPO database, elaboration by Angela Colucci.

worthwhile to notice that only 6.8% of the practices are specifically focused/targeted on actions themselves and that the great majority has a mix of projects goals, sectors involved, tools, and expected results.

The practices characterization have permitted to refine the database, leading to a further phase of selection to make the set of practices more homogeneous and comparable for the subsequent steps of analysis. The analyses carried out on the database refer to 134 practices (13 minus—8.8% of 147), since two typologies were not considered:

- Research (11 practices): these practices are not homogeneous and comparable with the others because, despite being research involving more actors, they have not activated local processes of transformation/innovation;
- Project/Proposal (2 practices): these practices are strongly focused on the development of a project or a project proposal that, however, have not (yet) found any implementation. These are interesting in reaching the result of building networks and strengthened local alliances but do not developed any implementation phase and, for this, are not comparable with other practices typologies.

5.3.3 Resilience practices: geographies and scales of action and benefits

The spatial dimension of practices constitutes a factor of analysis within two aspects: the "spatial domain" of action of practices and the "spatial domain" of benefits emerging/resulting from practices.

The project description of the practices presented by the promoters of the practices themselves when they joined RPO includes the spatial domain or the spatial area of actions/interventions (Tables 5.3 and 5.4).

Table 5.3 Resilience Practices Observatory research database: distribution according to the "spatial area" of action/intervention of practice.

Spatial dimension	Number	Percentage
Local	51	38
Municipal	38	28
Supramunicipal	26	19
Regional/territorial	17	13
National	1	1
Not applicable	1	1
Total	**134**	**100**

Bold value highlights the total line.
Credit: RPO database, elaboration by Angela Colucci.

Table 5.4 Resilience Practices Observatory research database: the "spatial area" of spillover/impact from the action/intervention activated by practices.

	Number	Percentage
Local	24	18
Municipal	39	29
Supramunicipal	37	28
Regional/territorial	31	23
National	2	1 (1.49)
Not applicable	1	1 (0.75)
Total	**134**	**100**

Bold value highlights the total line.
Credit: RPO database, elaboration by Angela Colucci.

The second aspect of analysis related to the "spatial domain" refers to the territorial basin of benefits resulting from the actions and interventions promoted by the practice.

The spatial area (spillover basin) of benefit resulting from actions conducted by practices shows the capacity in engaging different geographical scales or, at least, some specific components of external territorial context.

The categories describing the factor of "spatial" dimension are:

- local: the actions/interventions or the benefits are locally placed (e.g., strongly place-based, at the neighborhood level or strongly related to a local community);
- municipal level: the actions/interventions or the benefits are at the municipal level if the whole municipal communities and areas are involved/engaged;
- supramunicipal: the actions/interventions or the benefits involve more than a municipal area (e.g., the practices acting in a "rural district" or "rural intermunicipal park" involve different municipalities);

- regional/territorial: the actions/interventions or the benefits involve a territorial area (e.g., provincial district, territorial park, landscape components like river basin, or valley/mountain);
- national: only one practice (TCI—TerremotoCentroItalia) acts at the national level. The TCI is a no-profit and volunteer organization aimed to share information and opportunities to support (and organize) the communities of (Center) Italy affected by earthquakes or potentially at risk (https://terremotocentroitalia.info/);
- Not applicable: in setting the RPO database, the main goal was to connect the practices to a "spatial" domain. In the case of the "Sorgenti danza resilienza" practice (Dance Resilience Sources), the spatial dimension cannot be applicable: this association is national but organizes different very local-based workshops (boosting the individual/small communities resilience through the movement in the space; http://www.danzamovimentoterapia.it/dmt_italia.htm).

The data describe a situation based on local action, even very punctual one, with expectations in terms of positive externalities and imitative potentials, making the investments particularly effective from a territorial and systemic perspective. Only a few practices are conceived to remain in their small territorial frameworks, without an external dialog.

The observation of the data (number and percentages) of the distribution of practices based on the factor of spatial scale of benefits deriving from actions shows how practices mainly act at a local and municipal scale (in Table 5.3, respectively 38% and 28%); however, the practices provide benefits in larger scales. Concerning spatial scales of benefits, the practices are distributed in supramunicipal and territorial scales (19% and 12% in reading of scale of intervention and 27.61% and 23.13% in scale of impact). Despite having a local scale of action, the capacity of practices to activate relationships and therefore benefits at even higher spatial scales is well readable in Table 5.5, which shows in relation to each category of spatial scale of action the distribution of practices included in this category concerning the spatial scales of benefit. In particular, the table underlines how the 51 practices acting at local level demonstrated to provide benefits at local (13) but also at municipal (13), supramunicipal (7), and regional/territorial (7) levels and how the 38 practices acting at municipal level demonstrated to provide benefits at the municipal level (26) but also at supramunicipal (7), territorial (4), and national (1).

Table 5.5 Caption: Resilience Practices Observatory research database: connecting the "spatial area" of action and the spatial area of spillover/impact (the italicized values highlight the capacity of practices to positive impact higher spatial scales).

Spatial categories (row: of action; columns: of benefit)	Local	Mun.	SupraMun	Regional	Nat.	NA	Total
Local	24	*13*	7	7			**51**
Municipal		26	7	*4*	*1*		**38**
Supramunicipal			23	*3*			**26**
Regional/territorial				17			**17**
National					1		**1**
NA						1	**1**
Total	**24**	**39**	**37**	**31**	**2**	**1**	**134**

Bold value highlights the total line.
Mun, Municipal; *SupraMun*, supramunicipal; *Regional*, includes territorial scale includes territorial scale; *Nat.*, National; *Nat.*, National; *NA*, Not applicable.
Credit: RPO database, elaboration by Angela Colucci.

5.3.3.1 Geographical distribution

The RPO roots in Lombardy: the practices of the Resilient Communities grant of Cariplo Foundation have to be placed in Lombardy (plus the Verbano-Cusio-Ossola provincial district), and the RPFs were held in Milan. The two primary sources of practices engagement are strongly connected to the Lombardy regional context influencing practices engagement (Table 5.6). The practices are acting in Lombardy (64%), Piedmont (9%), and Emilia Romagna (7%). The concentration of practices located in Lombardy and northern Italy is due to the RPO engagement process. Based on this evidence, the RPO project activated initiatives to boost practices involvement from all the national territory (in the table, the "source" involving a higher number of practices outside the Lombardy region is the RPO source). However, at the same time, the action of the Cariplo Foundation in supporting the community-led initiatives (as well as other private foundations/actors and institutional actors) could have increased the activation of practices led by local/community actors in Lombardy (and northern Italy).

5.3.4 Resilience practices: the governance models

A large range of research and theoretical contributions highlight the relevance and the urgency of a renovation of governance of urban transformation processes to cope with the complexity of environmental, social, economic, and organizational phenomena characterizing the actual

Table 5.6 Resilience Practices Observatory research database: geographical region of action of resilience practices.

Region	RPO DATA BASE*		Source_FC CR call		Source_RP FORUM		Source_RPO	
	Number	Percentage	Num.	%	Num.	%	Num.	%
Campania	**2**	**1**		0	2	3		0
Emilia Romagna	**9**	**7**		0	3	5	6	38
Lazio	**4**	**3**		0	2	3	2	13
Liguria	**5**	**4**		0	5	8		0
Lombardy	**86**	**6**	*49*	*9*	*35*	*5*	2	13
Marche	**2**	**1**		0	1	2	1	6
Molise	**1**	**1**		0	1	2		0
Piedmont	**12**	**9**	5	9	4	6	3	19
Puglia	**1**	**1**		0	1	2		0
Sicilia	**1**	**1**		0		0	1	6
Toscana	**1**	**1**		0	1	2		0
Trentino Alto Adige	**1**	**1**		0	1	2		0
Veneto	**6**	**4**		0	5	8	1	6
International	**2**	**1**		0	2	3		0
Not applicable	**1**	**1**		0	1	2		0
Total	***134***	***100***	*54*	*100*	*64*	*100*	*16*	*100*

Bold value highlights the total line and the values of the complete RPO database.
FC CR call indicates practices engaged from the Cariplo Foundation "Resilient Communities" call; RP Forum indicates practices engaged during the RPF call; RPO indicates practices engaged during other activities of RPO.
Credit: RPO database, elaboration by Angela Colucci.

historical moment. The renovation of the governance process has to embrace cross-system interactions and multiscale phenomena and dynamics (temporal/spatial) of social–ecological systems (Ostrom, 1990, 2009). UN agencies launched campaigns promoting the implementation of "good urban governance" principles (for instance, the UN-HABITAT programs and the UNISDR campaign on Making Resilient Cities); at the same time, European interregional polices have focused on the renovation of governance approaches (Duit et al., 2010; International Strategy for Disaster Reduction & United Nations, 2012). The shared issue is the development of renovated governance models able to integrate emerging phenomena (bottom-up and local innovation initiatives) and environmental, social, and economic policies in the process of urban, metropolitan, and regional systems transformation (Lake, 2021; Salamon & Elliott, 2002).

Approaching the urban and territorial transformations from the "resilience practices" perspectives implies shifting the focus from the implementation of a tool or a single solution to the whole process, which goes from the partnership building through the design of the project up to its implementation. The processes therefore include all the activities characterizing not only the transformation/intervention of the projects but also all the practice implementation phases. These also imply the renovation of the toolboxes, according to the innovative environment which characterizes the resilience practices: innovation in knowledge (co)production, in decision-making process, in design and solutions identification (from "physical" and technical solutions to more complex and integrated strategies), implementation and management of initiatives emphasizing on the engagement and empowerment of all actors.

5.3.4.1 Actors and partnerships

RPO investigated the partnerships of practices. In the engagement process the practices declared the led partner or coordinator of the practice and the members of the partnership, information verified/integrated during the database finalization. Three are the central aspects of the investigation: the number of the partnerships members (main actors acting), the typology of coordinator actor, and the typology of the partnership members.

The partnership enables partners to have a more effective practice implementation and stabilization. In the RPO the partnership is a collaborative effort between two or more organizations sharing resources in acting the initiative toward common aims. The partnership is often flexible during the practice management, and it does not imply an equal distribution of power, resources, skills, and responsibilities. A broad array of arrangements characterizes the partnerships, from informal collaboration among associations to formal agreements. From literature emerge as the partnership among organizations is crucial in guaranteeing long-term stabilization and an advantage in multipurpose actions that sustain collective efforts to carry forward the development (Minkler et al., 2008). Moreover, the "fluidity" and heterogeneous situations emerging from the resilience practices partnerships analysis make challenging to assess the effectiveness and the relevance of partnership composition (due to the "intangible" or not explicit components). As Table 5.7 shows, small partnerships (from 1 to 5 partners) characterize 68% (91) of the resilience practices, the medium partnerships (from 6 to 10 partners) and large partnerships (from 10 partners) are equally represented 16% (respectively, 22 and 21 practices).

The analysis of the categories of actors coordinating practices has made it possible to integrate the understanding of the dynamics of partnerships. The typologies refer to the consolidated categories of actors usually assumed to analyze urban/territorial transformations (public or institutional actors; private/business actors and civil society). Since one of the fundamental themes and approaches of RPO is integrating scientific knowledge and research, research centers and universities are recognized as specific and investigated actor category. Civil society emerges as the category-leading the majority of resilience practices. The analysis examined the civil society detailing the associations and cooperatives actors' contribution.

Moreover referring to private sector actors, the analysis detailed the contribution of:

- private and profit actors (enterprises profit);
- the private foundations (often the foundations have no-profit mission but acted as private players);
- the rural consortia (that are association or corporation of private enterprises in general connected to farming/primary economic sector. These associations represent private interests even if they often act in the common interests and could be comparable in the experiences of the RPO to the actors of the Civil society;
- association and non-governmental organization (NGO) that represents 57% of practices.

The main coordinator actors are associations (57%), followed by public institutions (24%) and other actors from the research/academia shpere.

Table 5.7 Resilience Practices Observatory research database: coordinator partners categories of resilience practices.

Category (subcategories)	*RPO DATA BASE**		Source_FC CR call		Source_RP FORUM		Source_RPO	
	Number	Percentage	Number	Percentage	Number	Percentage	Number	Percentage
Civil society	***77***	***57***	**43**	**80**	**28**	**44**	**6**	**38**
Association	*59*	*44*	30	56	25	39	4	25
Cooperative	*18*	*13*	13	24	3	5	2	13
Research center/ university	***9***	*7*	**0**	**0**	**8**	**13**	**1**	**6**
Institutional body	***30***	***22***	**2**	**4**	**21**	**33**	**7**	**44**
Private actors	***18***	***13***	**9**	**17**	**7**	**11**	**2**	**13**
Private Foundation	7	*5*	6	11	0	0	1	6
Private actor/player	7	*5*	0	0	6	9	1	6
Rural consortium (private)	*4*	*3*	3	6	1	2	0	0
Total	***134***	*100*	**54**	**100**	**64**	**100**	**16**	**100**

Italic values highlight the numeric and percentage columns referring to the complete RPO database. The bold values indicates the total values of the macro-categories that aggregate the data subdivided into individual items.
FC CR call indicates practices engaged from the Cariplo Foundation "Resilient Communities" call; RP Forum indicates practices engaged during the RPF call; RPO indicates practices engaged during other activities of RPO.
Credit: RPO database, elaboration by Angela Colucci.

The relevance of civil society actors in coordinating the resilience practices is due to the number of funding programs (public and private calls) addressed specifically to the associations or the civil society actors and to the nature of resilience practices that are strictly connected to local communities' well-being and are often rooted on already consolidated experiences of spontaneous activities.

The partnership composition analysis (Table 5.8) shows how most practices are based on a partnership, including more than one category of actors. The 72% (97) of practices present a mixed partnership, including more than two actors from different typologies, and only the 28% of practices present a partnership composed of homogeneous actors (all coming from the same actors' typology as analyzed earlier).

Table 5.8 shows both the distribution for the "aggregated" results highlighting the number of practices characterized by homogeneous and mixed partnership and, at the same time, articulated for the main actors' typology composing the partnership. The partnerships among association and territorial/public actors are the most common among the partnerships based on mixed actors' representation.

Two unexpected phenomena emerge:

- the 11% of practice (15) have mixed partnerships engaging all the actors' categories, demonstrating a good level in involving significant stakeholders' representation in the practices management process;
- comparing the partnership composition with the number of actors involved in the partnerships (small, medium, and large partnership size), the emerging result confirms the capacities of grassroots initiatives in building alliances among different stakeholders: the majority of practices based on "small partnership" (with a maximum of 5 partners) demonstrates the ability to engage different typology of actors (the 66% of the small partnerships are composed by a mixed actors set).

5.3.4.2 The Governance processes

Most consolidated governance models at international scale are also recognizable in the RPO database:

- The institutional leadership characterizes the climate change adaptation and prevention risk programs and plans promoted by metropolitan/city government actors. The adaptation plans remain voluntary, strategic and are mainly characterized by an extensive range of participatory and inclusive paths to engaging various stakeholders and existing initiatives.

Table 5.8 Resilience Practices Observatory research database: partnerships composition of resilience practices.

Partnership	Number	Percentage	Small P (1–5 partners)		Medium (6–10 partners)		Large (> 10 partners)	
			Number	Percentage	Number	Percentage	Number	Percentage
Homogeneous partnership	***37***	***28***	***31***	***34***	***2***	***9***	***4***	***19***
H_Association	22	16	16	18	2	9	4	19
H_public/territorial actors	7	5	7	8		0		0
H _ private actors	4	3	4	4		0		0
H Research actors	4	3	4	4		0		0
Mixed Partnership	***97***	***72***	***60***	***66***	***20***	***91***	***17***	***81***
M_ among association and public/territorial	56	42	42	46	10	45	4	19
M_ among association and private	2	1	1	1		0	1	5
M_ among association and research	9	7	5	5	4	18		0
M_all categories represented	15	11	1	1	4	18	10	48
M_ among public/territorial and private	4	3	2	2		0	2	10
M_ among public/territorial and research	11	8	9	10	2	9		0
Total	**134**	**100**	**91**	//	**22**	//	**21**	//

Bold-italic value highlights the values of the macro-categories (sum) that aggregate the data of the categories listed above.
Credit: RPO database, elaboration by Angela Colucci.

- Bottom-up processes characterize transition initiatives that are activated, promoted, and implemented through the mobilization and actions of citizens. A key factor for the success of the projects, in terms of consolidation and continuity of initiatives, is setting up a solid and rich partnership around the initiative (with associations, NGOs, private sectors, and public/institutions).

The experiments of innovative governance models (Duit et al., 2010; Maesschalck, 2017; Olsson et al., 2006) are partially implemented and recognizable in the stakeholders' engagement and empowerment effort.

The significance of overcoming the typical dichotomy of categories top-down and bottom-up emerges from the experiences of RPO practices. Some "labeled" top-down practices related to climate change adaptation or risk prevention are characterized by a complex governance process able to engage in the decision-making process an extensive range of stakeholders. At the same time, investigating the nature of the governance process of community-led initiatives, it is crucial to properly evaluate the level of inclusivity of the bottom-up processes. Four possible scenarios summarize the RPO practices governance process:

- open process characterized by an institutional leadership,
- closed process characterized by an institutional leadership,
- open process characterized by non-institutional leadership,
- closed process characterized non-institutional leadership.

The openness and closeness of the governance process are evaluated on the engagement tools/initiatives declared and used in sharing the decision and the distribution of commitments/roles among the stakeholders involved. This initial governance processes exploration required a simplification of practices processes. In fact, to highlight commonalities, it was necessary to apply clusterization methods that forced and simplified situations that are fluid and dynamic (e.g., some community-led practices were characterized by an initial phase with a closed decision-making and governance process that then, in the process of upscaling, became inclusive engaging/empowering differentiated and more numerous actors). The simplification, however, demonstrated its efficacy in underlining original dynamics of governance processes overcoming the top-down and bottom-up categories (Table 5.9), showing how the practices are mainly adopting an open process for decision-making process both the institutional and non-institutional leads.

Table 5.9 Resilience Practices Observatory research database: governance process of resilience practices.

Governance model	Number	Percentage
Open process characterized by an institutional leadership	47	35
Closed process characterized by an institutional leadership	17	13
Open process characterized by noninstitutional leadership	51	38
Closed process characterized noninstitutional leadership	19	14
Total	**134**	**100**

Bold value highlights the total line.
Credit: RPO database, elaboration by Angela Colucci.

5.3.4.3 The stabilization and upscaling of processes

Two factors investigate the level of stabilization and upscaling of practices: the germination and the persistence over the time.

The persistence over time aims to examine if practices remain active (the evaluation is based on the presence of "news" published in the media and on the launch of new initiatives): 54% of practices are dynamic in proposing initiatives. Some of the practices (34.5%) are "closed" or have temporarily suspended the activities. These practices correspond substantially with initiatives launched and strongly linked to specific funds call/source (e.g., the Cariplo Foundation Resilient communities' program that requires an explicit period for actions implementation) or the ones oriented to a specific challenge or criticality to overcome.

The "germination" is a crucial aspect demonstrating the capacity of practices in activating a local process of social activation, including the capacity of upscaling. The purpose is to evaluate if practices activated in the local context can maintain and expand their scope and action, launching different initiatives connected and synergic to the central or original challenges.

Germination is a complex and challenging investigation aspect. It requires recognizing whether practices have triggered over time a broad process of innovation and social activation based on the multiplication and diversification of actions and initiatives, maintaining the recognizability of the process itself as a common "ground" connecting the different actions. The continuity (also partial) of actors involved in the partnership, the long-term goals and themes (challenges) of action, and the local context of intervention are the three factors assumed for idenfy the germination aspect. The activation of different paths of interventions originating from the core practices was investigated; in relation to each new path (if registered), the prerequisites in considering this as germination were verified (actors' partnerships, action and benefits context, challenges, and long-term goals). The 47% of practices germinated other initiatives synergic with the initial one, and the 9% of practices launched secondary initiatives (evaluated as partial germination because the initiatives found are characterized for local and short-time actions). The 36% of practice does not present upscaling and germination (partially overlay with the practices "closed" above cited). Finally, as for the "stabilization", it was not possible to reach a confident evaluation of the germination factor in the case of the 7.4% practices.

5.3.4.4 Beneficiaries and targets of resilience practices

The majority of practices present multiple beneficiaries because these often are activating multi-issue actions. Representative categories of targets describe the rich and articulated panorama of beneficiaries synthetically. In presenting the results of this evaluation aspect, it is necessary to make two comments. This synthesis considers only the direct targets declared by practices assumed as to whom the activated actions were addressed as a priority. Indirect or secondary beneficiaries constitute an interesting but complex aspect of being evaluated because information on that is fragmentary and often overlaps with direct targets information. The incomplete data derive also from the missing information and the uncertain identification of secondary beneficiaries by the practice coordinators during the RPO data collection.

The data exploration (Table 5.10) confirms that the majority of practices is addressed to a generic target of "citizens/communities" (67%). The identification of a "generic" beneficiaries/target confirms the registered difficulties in the clearly/aware identification of direct and indirect targets by the coordinators/actors of practices.

An interesting aspect that emerges is that a good number of practices have as their direct and primary beneficiaries economic actors and, more specifically, agricultural producers/local businesses (16% agricultural producers/local businesses and 1% local businesses), while the 6% of practices address to local authorities. The target of the 8% of practices is vulnerable/fragile/young people.

Table 5.10 Resilience Practices Observatory research database: direct target categories of resilience practices.

	Number	Percentage
Local community/citizens	90	67
Vulnerable/fragile/young people	11	8
Local authorities	8	6
Experts/research	2	1
Local farmers/industries	22	16
Local businesses	1	1
Total	**134**	**100**

Bold value highlights the total line.
Credit: RPO database, elaboration by Angela Colucci.

5.3.5 Criticalities, topics, and tools

This section presents three aspects investigating the criticalities, core topics, and the *tools domains* of resilience practices. The aim is to understand and describe synthetically the geographies of core topics and urgencies faced by the practices. An incredible complexity characterizes the resilience practices concerning issues addressed and, multiple levers of actions could be recognized over the process, demonstrating a dynamism in adapting tools issues according to the local situation end needs evolution. Practices that shifted the attention and efforts in the quality of the process itself (more than in the interventions or single solution) have demonstrated to approach the solution design in an incremental approach, reorienting the single solutions in the practice process.

In particular, the three aspects were investigated:

- the urgencies from which the practices move or the lever that activated the action,
- the key topic around which they act, and
- the tools used to achieve the goals and overcome the urgencies addressed.

The urgencies, key topic, and tool domain are clustered concerning some general key concepts. Some keywords or key concepts are the same and recur in the analysis of three aspects, but rarely the same key concept is repeated at the single practice scale. A single practice rarely identifies identical/similar key concepts referring to urgencies, key topics, and tool domains.

5.3.5.1 The critical issues: levers for activation

The form of RPO adhesion includes a section related to the criticalities or urgencies in which the practice addresses explicitly. The criticalities identified by practices have been associated/reduced to ten categories emerging from a clustering process involving the practices and based on dialogue/feedback and co-production activities.

Through a further forced grouping (Table 5.11), it emerges that 48% of practices are motivated by urgencies connected with environmental issues [divided into loss of habitat and biodiversity, loss of agricultural quality, environmental pressures (cc/risks/CS), territorial risks, hydrogeological risks/disasters]. The 39% of practices start from urban/territorial crisis or shrinking (composed of shrinking of activities and territories, economic crisis and from shrinking of internal areas and urban degradation) and the 13% from the social fragilities both at the individual (individual resilience, few practices) and the community level (loss of social cohesion/fragility). Regarding the single categories, the shrinking of activities and territories and the loss of habitat and biodiversity emerge as the two most common critical phenomena to face.

Table 5.11 Resilience Practices Observatory research database: criticalities addressed by resilience practices.

	Number	Number	Percentage	Percentage
Individual resilience	4		3	
Loss of social cohesion/ fragility	14		10	
Social		***18***		***13***
Shrinking of activities and territories	31		23	
Economic crisis/fragility	7		5	
Urban degradation	14		10	
Urban/territorial crisis/ shrinking		***52***		***39***
Loss of habitat and biodiversity	20		15	
Loss of agricultural quality	4		3	
Environmental pressures (cc/risks/CS)	24		18	
Territorial risks	2		1	
Hydrogeological risks/ disasters	14		10	
Environmental		***64***		***48***
Total	**134**	**134**	**100**	**100**

Bold-italicized rows highlight the total concerning the macro-categories (single categories clustered follow).
Credit: RPO database, elaboration by Angela Colucci.

5.3.5.2 Core-topic clusters

A process of clustering permitted to understand the geographies of core topics is addressed by resilience practices. The first identification of key topics addressed arises from the analysis of the narrative contents of the database (in particular, the section describing the vision/goal and the one presenting principal strategies activated). The key concepts emerging were reorganized/grouped in nine clusters representing the core topics addressed by practices. The nine clusters emerging from the database analysis are consistent and aligned with the principal topics of interest of resilience action that emerged during the coproduction and codesign activities of the RPO (Table 5.12).

The majority (33%) of resilience practices focus on regeneration visioning and intervention: practices mainly regard projects and solutions for urban regeneration (urban peripheries and peri-urban areas in metropolitan contexts) and the sustainable reactivation of internal areas (urban and local regeneration 19%, local development 14%). Climate change mitigation and adaptation (13% of practices) and territorial risks (7% of practices) represent the core-topic clusters for the 20% of resilience practices. The practices addressing climate change adaptation topics are acting at regional or territorial scales and are usually related to institutional processes already activated (e.g., Mayor adapts program or energy transition policies). Practices on sustainable river basin management (characterized by an extensive partnership including institutions and national associations) include policies and measures for risk prevention. Few of these focus on

Table 5.12 Resilience Practices Observatory research database: core-topic domain of actions of resilience practices.

Topic domain (subtopic)	Number	Percentage
Institutional innovation	6	4
Social innovation	13	10
Individual resilience	3	2
Ecosystem services/biodiversity	14	10
Territorial risks	9	7
Climate change	17	13
Agrifood supply chains	28	21
Urban and local regeneration	25	19
Local development	19	14
Total	**134**	**100**

Bold value highlights the total line.
Credit: RPO database, elaboration by Angela Colucci.

communities' risk preparedness and prevention regarding self-help preparation and intervention to reduce building vulnerability.

The 21% of practices have in the food chains cluster the core topic. In all the practices, the food topic is approached in terms of sustainable food system development. In most mapped practices the food topic is strictly connected to natural and ecosystem preservation (ecosystem services, natural areas improvement, agroecology).

Practices focusing directly and specifically on natural resources cycles, including biodiversity and natural habitats improvement, are the 10% of practices.

Practices addressing as core focus on social innovation (10%) and institutional innovation (4%) are also represented. In particular, social innovation is mainly faced/approached as a "tool" for facing urgencies more than identified as a core topic. The fact that social innovation and social inclusion aspects have a more limited representation as a core topic could also be partially ascribable to the large presence of Cariplo Foundation practices. The Cariplo Foundation Resilient Communities program focused on environmental and territorial risks, promoting the local communities' capacities to face climate change and territorial/environmental risks and boosted Resilience practices in addressing these two components as core topics. At the same time, several supported practices by Cariplo Foundation were focused on the food system as a core topic in overcoming local criticalities (for instance, several practices face the hydrogeological risks in internal areas activating or reactivating local agrifood chains able to develop local economies and guarantee the maintenance of internal areas).

5.3.5.3 Tools for action domains

The practices, as described previously, have activated and managed complex processes articulated on multiple initiatives and actions.

To highlight/make evident the principal tools used to achieve the practice goals, it was necessary to reduce this complexity and richness. In particular, the simplification of the richness and complexity affected:

- the reduction of the innovative and creative tools, ideas proposed and acted in the practices' process to seven clusters of action-tool domains (capacity building, codesign, institutional innovation, social innovation, direct intervention, new economies, and knowledge framework);
- the timeframe or the phase of single tool action activation: each practice is often activating several parallel activities and initiatives using

different instruments or tool domains. In this analysis the identification (and assignation) of the principal tool domain characterizing each practice is based on the already launched/implemented action (not necessarily completed but at least launched), and it has not considered the planned actions;

- the scale of impact: the scale impact is another additional criterion in estimating the tool action domain. In cases of multiple initiatives activated and referable to different tool action domains, the category was selected/assigned in reason to the activity (and related action-tool domain), providing the most significant scale of benefit. For instance, often physical interventions are characterized in having local (local-based and local-placed) direct impacts, but all the activities promoted along the codesign and coproduction process for the "physical" intervention realization provide a large range of urban/territorial scale benefit in boosting the capacities and the awareness of local communities and local/supralocal actors.

Table 5.13 shows that the capacity building tool domain (32%) represents the most adopted, or at least the most implemented/launched, tool in the first steps of the practices' processes. The first phases emerged as phases devoted to improving the capacities of the association/partners managing the practices and local communities or specific target of beneficiaries. For example, several practices on sustainable food chain were launched in the first phases initiatives to improve the capacities addressed both citizen-consumers (and their behavior) and farmers' capacities.

Similar considerations could be referred to the two other action-tool domains very often activated by practices: tools for knowledge improvement

Table 5.13 Resilience Practices Observatory research database: tool domains of actions of resilience practices.

Tool domain	Number	Percentage
Capacity building	43	32
Codesign	15	11
Institutional innovation	3	2
Social innovation	8	6
Direct intervention	17	13
New economies	24	18
Knowledge framework	24	18
Total	**134**	**100**

Bold value highlights the total line.
Credit: RPO database, elaboration by Angela Colucci.

(18%) and tools for codesign (11%). This result shows the crucial role also recognized by resilience practices to the "process" of transition. The three action-tool domains mostly adopted are tools for improving communities and actors' knowledge and capacities and managing local actor empowerment. These three action tools demonstrate both the shifting of attention from physical/spatial intervention to the process itself and the recognition (awareness) of the crucial role played by the design of the governance aspects of grassroots practices.

The tools devoted to spatial or immaterial transformation interventions (direct intervention 13%, activating new economies 18%, social innovation 6%) are less represented. Among the intervention tool domains, the new economic model (new economies) is the most represented. The activation of the tool in implementing and launching new economic models (e.g., sustainable/social enterprises, reactivation of local/sustainable food chain, activation of local circular economies) is a relevant (and unexpected) result demonstrating the knowledgeable/aware orientation of resilience practices for their stabilization and the ability to provide tools and activities to sustain the sustainable local long-term development.

Acknowledgments and Attribution

The authors are grateful to all the components of the Resilience Practices Observatory project as well as to all the practices actors engaged during the RPO activities.

Although the chapter is the result of a common reflection, Sections 5.2 and 5.3 have been written by Angela Colucci and Section 5.1 is the result of common contribution.

References

Berkes, F., Folke, C., & Colding, J. (2000). *Linking Social and Ecological Systems: Management Practices and Social Mechanisms for Building Resilience*. Cambridge University Press.

Burayidi, M.A., Allen, A., Twigg, J., & Wamsler, C. (2020). *The Routledge handbook of urban resilience*.

Colucci, A. (2012). Towards resilient cities. Comparing approaches/strategies. *Tema. Journal of Land Use, Mobility and Environment*, 101–116. Available from https://doi.org/10.6092/1970-9870/921, Paginazione.

Colucci, A. (2018). *Transition initiatives: Three exploration paths. Smart, Resilient and Transition Cities: Emerging Approaches and Tools for a Climate-Sensitive Urban Development*. Elsevier.

Duit, A., Galaz, V., Eckerberg, K., & Ebbesson, J. (2010). Governance, complexity, and resilience. *Global Environmental Change*, *20*(3), 363–368. Available from https://doi.org/10.1016/j.gloenvcha.2010.04.006.

Galderisi, A., & Colucci, A. (2018). *Smart, Resilient and Transition Cities: Emerging Approaches and Tools for A Climate-Sensitive Urban Development*. Elsevier.

International Strategy for Disaster Reduction & United Nations. (2012). *How to make cities more resilient: a handbook for local government leaders. A Contribution to the Global Campaign 2010-2015, Making Cities Resilient-My City is Getting Ready*! United Nations.

Lake, D. A. (2021). The organizational ecology of global governance. *European Journal of International Relations*, *27*(2), 345–368. Available from https://doi.org/10.1177/1354066120959407.

Maesschalck, M. (2017). *Reflexive Governance for Research and Innovative Knowledge*. John Wiley & Sons, Incorporated. Available from http://ebookcentral.proquest.com/lib/polimi/detail.action?docID = 4816329.

Minkler, M., Vásquez, V. B., Tajik, M., & Petersen, D. (2008). Promoting environmental justice through community-based participatory research: The role of community and partnership capacity. *Health Education and Behavior*, *35*(1), 119–137. Available from https://doi.org/10.1177/1090198106287692.

Olsson, P., Gunderson, L. H., Carpenter, S. R., Ryan, P., Lebel, L., Folke, C., & Holling, C. S. (2006). Shooting the rapids: Navigating transitions to adaptive governance of social-ecological systems. *Ecology and Society*, *11*(1), 18.

Ostrom, E. (1990). *Governing the Commons: The Evolution of Institutions for Collective Action* (1st ed.). Cambridge University Press. Available from https://doi.org/10.1017/CBO9780511807763.

Ostrom, E. (2009). *Understanding Institutional Diversity*. Princeton University Press.

Roostaie, S., Nawari, N., & Kibert, C. J. (2019). Sustainability and resilience: A review of definitions, relationships, and their integration into a combined building assessment framework. *Building and Environment*, *154*, 132–144. Available from https://doi.org/10.1016/j.buildenv.2019.02.042.

Salamon, L. M., & Elliott, O. V. (Eds.), (2002). *The Tools of Government: A Guide to the New Governance*. Oxford University Press.

CHAPTER 6

Monitoring the Fondazione Cariplo Resilient Communities practices

Marcello Magoni and Rachele Radaelli
Climate Change, Risk and Resilience Laboratory, Department of Architecture and Urban Studies (DAStU), Politecnico di Milano, Milan, Italy

6.1 Introduction

The monitoring of an implementation process of actions and interventions envisaged by a project or a plan requires two distinct conceptually steps to be carried out in an integrated manner. The first concerns the identification and measurement of the indicators that are considered to represent the process to be monitored in an optimal way. The second concerns the evaluation of these measurements with respect to criteria representative of the evaluation purpose, which is generally constituted by the objectives of the actions. When monitoring concerns practices that are aimed at increasing the resilience in territorial systems, then these steps must be developed on the basis of a rigorous definition of resilience, which should be made according to the type of evaluation that is intended to be performed and the object of the practice.

In this case the evaluation was carried out to know the effects and magnitude of the resilience improvement in the territorial systems affected by the resilience practices financed by the Fondazione Cariplo. In particular, through the call named *Resilient Communities*, Fondazione Cariplo financed 54 practices over the course of 4 years, of which 9 were evaluated because they were the only ones completed by the end of 2018, when the monitoring activity began.

Given that no evaluation method functional to evaluate the effectiveness of an action in improving the resilience of a territorial system was available in the literature (Carpenter et al., 2001; Chelleri, 2015), a method suitable to evaluate that kind of effectiveness was then developed. Therefore the description of the characteristics of this method and the

[ECO]systems of Resilience Practices
DOI: https://doi.org/10.1016/B978-0-12-819198-9.00011-9

relative criteria and evaluation factors is a necessary step to understand the evaluations results of those practices.

This essay is divided into two parts.

The first part describes the method and the related evaluation characters, criteria, and factors used to monitor what has been carried out by resilience practices and to evaluate their effectiveness.

The second part describes the characteristics of the analyzed practices, the main evaluation passages, and the relative results. Furthermore, some general considerations are formulated that can be modified in the future with the evaluation of all 54 resilience practices financed by the Fondazione Cariplo. In this regard a limitation to the generalization of these assessments is due to the fact that these practices were designed to respond to a call that led them toward the resolution of territorial, social, economic, and/or environmental problems.

6.2 The evaluation of resilient practices for territorial systems

The literature on territorial resilience has been the subject of different elaborations by numerous scholars of several disciplines (Davoudi et al., 2012; Olsson et al., 2015; Fleming & Ledogar, 2008; Pendall et al., 2010; Reghezza-Zitt et al., 2012; Walker & Salt, 2012), thus becoming a sort of "tapestry of definitions and meanings with little orthodoxy in its conceptualization and application" (Cutter, 2016). If, in general, resilience can be defined as a property of a territorial system that allows it to survive and thrive in situations of uncertainty, adversity, and change, to use its definition as a reference to develop an evaluation method, it is necessary to clarify the concept to adopt and its underlying principles.

To do this, it was decided to take into account the broader range of indications contained in the numerous approaches to resilience that have been proposed in the international literature by referring to the three strands proposed by Martin and Sunley, (2015), which are distinct by the prevailing capacity of a territorial system to oppose, reconstitute itself, or evolve. Among these approaches the one indicated with the term of coevolutionary resilience is of considerable interest (Davoudi et al., 2013; Folke et al., 2010; Holling, 1973; MacKinnon & Derickson, 2013). It considers the resilience as a constantly evolving process that seeks to transform crises into development opportunities, for which the resilience is not only seen as a capacity to recover from natural and/or anthropic disasters, but also as the long-term evolution of the territorial system itself. This capacity implies the sharing of change and therefore the predisposition for a territorial system to react to potential shocks and stresses through the realization of "quality leaps" that are useful to anticipate the changes. To encourage this sharing, it is necessary to act on those values of the community that give stability to the territorial system and that can take place by regenerating, with the broadest consensus, the identity, and symbolic references of the community (Portugali, 2000).

The resilience of a territorial system is considered here as its ability to mitigate the impact of shocks and stresses on its structures and infrastructures and on the health of its community, to continue to function or to be able to quickly restore its essential functions and adapt to reduce interruptions from future events. Furthermore, are considered the context and the territorial scale, the phases of the disaster management cycle (mitigation, preparation, absorption, recovery, response, and adaptation), and the assumption that the resilience practice aims at improving the quality itself of the evaluation process (Sharifi et al., 2017).

Being in support of a monitoring activity, this method was used in this case to carry out an ex post evaluation, that is, an evaluation following the realization of the considered practices. It can also be used to carry out an ex ante evaluation, that is, to support the process of elaboration of the resilience practices. In this second case, it would become very useful to evaluate in an integrated way the different elements of the territorial system, maximizing their synergies and minimizing their trade-offs (Olazabal et al., 2018; Sharifi & Yamagata, 2018).

6.2.1 The theoretical and methodological characters

The evaluation method developed falls within the family of multidimensional methods (Fusco Girard, 2011) and adopts a systemic approach which, in the evaluation of broad and articulated strategies and practices, requires to be developed in two phases: in the first phase the conceptual scheme of the territorial system of the considered practice is built, and in the second phase the evaluation of the practice is carried out on the basis of the established evaluation factors and application criteria.

The conceptual scheme of the territorial system is built through the following steps:

1. identification of the spatial, temporal, dimensional, and actorial boundaries of the *reference system*, which constitutes the part of the territorial system directly affected by the practice to be evaluated;

2. identification of the components of the *reference system* and their internal and with the upper and lower level systems' interactions;
3. identification of the dynamics of the *reference system* in the absence and presence of the planned interventions.

Since the resilience practices considered have a very limited impact on the territorial system, their evaluation did not make necessary to develop this first phase in depth because it was enough to consider in an expeditious way the possible effects on the higher level systems.

The factors identified for the resilience practices evaluation and the related application criteria have taken into account the definition of territorial resilience given previously and the relative approach. They are *adaptability, robustness, redundancy, diversity, connectivity, social cohesion, information/knowledge, awareness/public spirit/self-organization.*

The *adaptability* and *robustness* are defined as *primary resilience assessment factors* because they are considered higher level than the other factors, which are defined as *secondary resilience assessment factors*. This is because the resilience of a territorial system is a function above all of the ability to adapt and the availability of resources (Longstaff et al., 2010), the latter considered in a broad sense in which economic, human, organizational, cognitive, and infrastructural resources are included. In turn, *primary resilience assessment factors* are partly a function of *secondary resilience assessment factors*.

Adaptability is the capacity of a territorial system to dynamically conform to its environment when it has an insufficient or inappropriate amount of resources necessary for its survival and development. This capacity requires an active attitude by the system, through the implementation of strategies and actions that aim to overcome the critical issues by identifying and achieving long-term objectives. The *robustness* of a territorial system is given by a sufficient availability of reliable, stable, and integrated resources, capable of supporting the functioning and performance of the system with respect to the critical issues and/or objectives that a strategy or an action should, respectively, counter or achieve.

To avoid evaluating the same property more times, only those properties that are not analyzed in *secondary resilience assessment factors* are considered in the evaluation of *adaptability* and *robustness*. Among these properties, there are also "emergent properties," which are the consequence of the interactions between the different factors that characterize a system and which can be identified and evaluated on a higher logical scale that are represented by *adaptability* and *robustness*.

Redundancy is considered here as the existence of a quantity of resources greater than those necessary to perform a function, to ensure that this function can continue to be performed even in the event of insufficient availability of one or more of those resources (Chuvarajan et al., 2006). *Redundancy* affects *adaptability*, since it allows for a greater variety of available resources, and *robustness*, since it increases the consistency and stability of resources and their performance. It is complementary to *diversity* and *connectivity* in giving stability and security to a system, thus indirectly favoring *social cohesion*.

The increase in redundancy can compromise efficiency in the functioning of a system and this effect, especially in the growth phase of the system, can reduce its competitiveness. On the other hand, redundancy is often very useful in the critical phases of the adaptive cycle of a territorial system (Marcus & Colding, 2014). From this point of view, to identify the most effective and appropriate solutions, it is necessary to deeply know the interrelations between the various factors and to pay attention to the multiple dimensions of resilience.

Diversity implies the existence in a territorial system of a quantity of different resources that carry out the same functions and that can be overall higher than those required in the ordinary periods of operation, so that the system can continue to perform its functions in an acceptable way in the event of an available resources reduction.

The existence of a good *connectivity* between the elements of a territorial system and between the latter and other systems makes it possible to increase the capacity of the system to counteract or absorb shocks and stresses. *Connectivity* influences *adaptability*, as it makes available more flexible resources, and *robustness*, since it influences the consistency and stability of resources and their performance over time.

Social cohesion represents the set of behaviors and bonds of affinity, trust, and solidarity among the individuals of a community, behaviors that are oriented to mitigate the disparities of social, economic, cultural, and ethnic conditions. *Social cohesion* influences the *robustness* of a community as it increases its capacity to respond and interact with *knowledge, awareness*, and *public spirits* in its ability to give proactive and innovative answers.

In a territorial system, *information* and *knowledge* develop and reinforce each other and both these factors form the basis of a profitable relationship with *awareness, public spirit*, and *self-organization*. The *information* and *knowledge* influence the *adaptability*, as they favor the timeliness of the community's and institutions' reaction, and the *robustness*, since they favor the achievement of solid and harmonious social relations.

The *information/knowledge, awareness/public spirit/self-organization* mutually develop and reinforce each other and contribute to build communities aware of their vulnerabilities and capable of acting proactively. They influence the *adaptability*, as they favor the timeliness of the actions of a community and its institutions, and the *robustness*, since they positively influence social and institutional connections and have synergies with *information/knowledge* and *social cohesion* in giving proactive and innovative answers.

This evaluation method should consider the hierarchical relationships starting from those of lower level to finish to those of higher one. Therefore the individual *secondary resilience assessment factors* and then the corresponding reports are initially evaluated. The two *primary resilience assessment factors* are then evaluated individually, considering the respective relationships with the lower level factors. Finally, *resilience* is evaluated on the basis of the relationships that exist between the two *primary resilience assessment factors* and taking into account the characters of the *secondary* ones.

6.2.2 The application characters

The evaluation of resilience practices is carried out considering both their potential effect, which concerns the planned actions and it is functional to understand the effectiveness of each practice with respect to its goals regardless of their being directed toward conditions of higher resilience, than the real one, which concerns the actions carried out. In the first case, between the objectives and the relative actions, the coherence is verified and the consistency is assessed, and in the second case the monitoring of what has been achieved with a practice, the verification of the achievement of objectives, and the evaluation of the resilient capacity improvement of the system are made.

The verification of the coherence is carried out by building, on the basis of explicit indications of the projects and, where necessary and possible, integrated by deductive considerations, a framework composed of three elements: *objectives, goals, actions*. The coherence assessment is then carried out first between every *objective* and the relative *goals* and then between the latters and the planned *actions*. The rating scale used consists of five classes: inconsistent (INC), indifferent (IND), low (L), medium (M), high (H).

The evaluation of the consistency is done only for the *actions* that are consistent with the *goals* and it involves the evaluation first of the consistency between *actions* and *goals* and then between the *goals* and *objectives*. The evaluation is done through a scale of five classes which, being used for most of the evaluations made, is here defined as *standard rating scale*. Those classes are low (L), medium-low (ML), medium (M), medium-high (MH), high (H). When the specific evaluation criteria are not indicated, the general criterion is adopted so that one starts by attributing to the considered factors the high, medium, or low rating depending on the performances found. If the assigned *rating* is not considered satisfactory, the MH or ML intermediate ratings are used. For consistency, a low rating is assigned to the *actions* that, due to their size and significance, have a low capacity to achieve the relative *goal* and *objective* (figuratively less than 20%); a medium-low rating to the *actions*, the capacity of which is mediocre (figuratively between 20% and 40%); a medium rating to the *actions*, the capacity of which is fair (figuratively between 40% and 60%); a medium-high rating to the *actions*, the capacity of which is significant (figuratively between 60% and 80%); a high rating to the *actions* that in terms of number and significance are potentially capable of achieving the related *goal* and *objective* (figuratively over 80%).

The evaluation of the real effect of a practice is articulated in two steps: the evaluation of the correspondence between the planned *actions* and the *actions* done and the evaluation of the level of achievement of the *objectives*. Both these assessments are made using the *standard rating scale*.

In the first case the following criteria are applied: a low correspondence is attributed when the set of actions done covers up a part lower than 20% of the planned *actions*, taking into account the different capacity to impact the system of the latters; an ML correspondence is attributed when the set of actions done involves a part between 20% and 40% of those planned; a medium correspondence is attributed when the set of actions done involves a part between 40% and 60% of those planned; an MH correspondence is attributed when the set of actions carried out involves a part between 60% and 80% of those planned; a high level of correspondence is attributed when the set of actions done involves more than 80% of those planned.

In the second case the following criteria are applied: a low correspondence is attributed when the *actions* carry out a minor part of the relative *objective* (figuratively less than 40%), a medium correspondence when the *actions* carry out a significant part of the relative *objective* (figuratively between 40% and 70%), a high correspondence when the *actions* achieve many parts of the relative *objective* (figuratively over 70%).

The evaluation of the improvement of resilience capacity is done on the basis of the factors and evaluation criteria described in the previous chapter, comparing the actions with the type and severity of the criticalities to contrast. In particular, taking into account the specific characteristics and the related interactions in achieving the required performances, the factors *social cohesion, information/knowledge*, and *awareness/public spirit/ self-organization* are initially evaluated, while subsequently are evaluated the factors *redundancy, diversity*, and *connectivity*.

In both cases the evaluation is made in two steps: in the first one, it is assessed whether the practice has significantly changed, both positively and negatively, the evaluation factor considered. If so, the assessment is carried out using two symmetric (positive and negative) *standard rating scales*, otherwise, the evaluation finishes and it is attributed to the factor considered the irrelevant change *rating*.

Using the *standard rating scale*, the *adaptability* and *robustness* factors are then assessed individually, taking into account their specific characteristics and the assessments given to the *secondary resilience assessment factors*. The final evaluation on resilience improving *reference system* is made on the basis of the assessments given to the *adaptability* and *robustness* factors and considering the assessments given to the *secondary resilience assessment factors*.

6.3 Outcomes and considerations of resilient practices monitoring

The nine monitored resilience practices were carried out in the northwest of Italy (see Fig. 6.1 and Table 6.1) and implemented between 2015 and

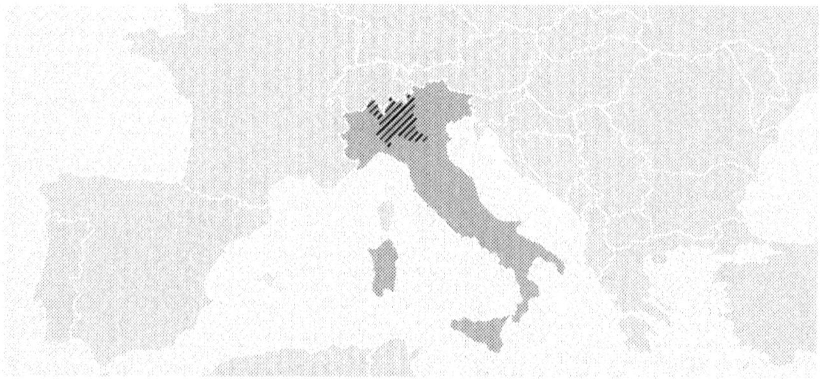

Figure 6.1 Area in which the set of monitored resilience practices was implemented.

Table 6.1 List of monitored resilience practices.

Eco-Food Chain/Filiera Eco-Alimentare Valorization of game meat Leader: Associazione per Sviluppo Cultura, Studi Universitari e Ricerca nel Verbano Cusio Ossola http://www.osservatorioresilienza.it/progetti/comunita-resilienti/2014/1451
Three Parks Chain/Tre parchi in filiera Experimentation of an agricultural development model Leader: Fondazione Idra http://www.osservatorioresilienza.it/progetti/comunita-resilienti/2014/1487
The Blond Goat/La Capra Bionda Valorization of native fauna in mountain environment Leader: GAL Vallecamonica Val di Scalve http://www.osservatorioresilienza.it/progetti/comunita-resilienti/2014/1491
GE.COO.FOR. Coordinated Management of Forests Leader: Consorzio Forestale Lario Intelvese http://www.osservatorioresilienza.it/progetti/comunita-resilienti/2014/1503
Not only Lake/Non solo lago Tourist-environmental valorization in the mountain area Leader: Istituto Oikos Onlus http://www.osservatorioresilienza.it/progetti/comunita-resilienti/2014/1513
Resilient Gardens in Carate Brianza/Orti resilienti di Carate Brianza Recovery of degraded areas for urban gardens Leader: Demetra Società Cooperativa Sociale Onlus http://www.osservatorioresilienza.it/progetti/comunita-resilienti/2014/1555
The Lombard System of Participative Guarantee/Il Sistema Lombardo di Garanzia Partecipativa Quality certification of the food chain Leader: Associazione Italiana per l'Agricoltura Biologica Lombardia http://www.osservatorioresilienza.it/progetti/comunita-resilienti/2014/1557
The Door of the Park/La Porta del Parco s.n.c. Participated recovery of a degraded area Leader: Alchimia Società Cooperativa Sociale http://www.osservatorioresilienza.it/progetti/comunita-resilienti/2014/1560
Crazy-lienza Realization of a Local Resilience Centre Leader: Liberi Sogni Cooperativa Sociale http://www.osservatorioresilienza.it/progetti/comunita-resilienti/2014/1566

2018. All the practices lasted between 18 and 24 months and they have all been completed satisfactorily, even if with different levels of completeness with respect to the stated objectives.

6.3.1 The characteristics of resilience practices

The practices are located in about half of the cases in mountain areas (Eco-Food Chain, The Blond Goat, GE.COO.FOR., Not only Lake); in two cases, they are situated in agricultural peri-urban areas (Three Parks Chain and Participative Guarantee) and in the remaining three cases, they are located in abandoned and/or degraded peri-urban areas (abandoned woods or vineyards, unused areas, etc.).

As required by the call *Resilient Communities*, the projects are promoted by nonprofit organizations, the heads of which belong, in almost all cases, to the third sector (social cooperatives or nonprofit associations) and, in only one case, to an institution constituted by a forestry consortium. The partnership networks are wide and involve, with different geometries, the associations (environmental, social, etc.), the economic operators (producers, breeders, tourist agencies, trade associations, etc.), and, in a smaller number of cases, local institutions (municipalities, unions of municipalities, park authorities and mountain communities, etc.), schools, and universities. The two practices that involve administrative bodies and universities aim at verifying the feasibility of experimental models of resource management and pilot projects (Eco-Food Chain and Participative Guarantee), while the other practices that involve associations, schools, and communities aim at stimulating new forms of bottom-up local resources use and fruition (Three Parks Chain, Resilient Gardens, The Door of the Park, and Crazy-lienza) or at involving the socioeconomic operators interested in solving the existing critical issues in some economic sectors (The Blond Goat, Not only Lake, GE.COO.FOR.).

The most recurrent critical issue, which characterizes all practices even if in different forms and intensities, is the loss of territorial quality. In particular, more than half of the practices deal with the problems of abandoning activities and territories at the same time as the hydrogeological risk (see, e.g., GE.COO.FOR. and Crazy-lienza) or the habitat and biodiversity loss (The Door of the Park and The Blond Goat) or both (Not only Lake), while the remaining practices consider relevant the loss of agricultural quality (such as the Eco-Food Chain and Participatory Guarantee) and the risk of land consumption (Three Parks Chain and Resilient Gardens).

The most frequent objective that the practices intend to achieve is the prevention or reduction of environmental and territorial risks. In most cases, this objective is combined with the enhancement of the territory and the collective care and sustainable management of territorial resources (The Door of the Park, Resilient Gardens, GE.COO.FOR.). The promotion of sustainable agricultural systems and the support for proximity economies are the objectives of the remaining practices (Three Parks Chain, Participatory Guarantee, and The Blond Goat).

6.3.2 The assessment of the potential and real effectiveness of resilience practices

The totality of the monitored practices has a good level of potential effectiveness, due to a correct structuring of the relations between the problems encountered, the *objectives* they intend to achieve, and the necessary actions for achieving them. Only for three cases (GE.COO.FOR., Participative Guarantee, and Crazy-lienza), it was necessary to make, in the analysis phase, some logical-deductive integrations and modifications to better explicit the *objectives*, the actions, or their relations, having been implicitly indicated in the project proposals.

In all the practices the actions have a high coherence with the reference *objectives*, but only in three of them, they have a high consistency. In other practices, actions are scarce by number or not fully adequate to achieve one or more objectives. In particular, in half of the cases, there are no specific actions relating to the involvement of some categories of stakeholders that have a key role both for identifying local problems and needs and for carrying out the actions (Three Parks Chain, GE.COO. FOR., Not only Lake); in other cases, there are no organizational and managerial actions that guarantee the involvement of stakeholders, which are an essential factor for the practices feasibility (The Door of the Park and Crazy-lienza).

An outcome noted in all the practices is the considerable gap existing between their *aims*, which consist of activating *responses* to stem or reverse serious territorial, environmental, and social criticalities and/or to modify or innovate behaviors and resources production, use and consumption models, and the *objectives* that, realistically, the set of planned actions in each practice allows to achieve.

The planned actions have been almost entirely done in all the cases, with the exception of some of them where some problems have affected the completion of the actions or they have been replaced with more feasible actions

The difficulties in carrying out the actions are in almost all cases due to problems in the involvement of project partners and/or stakeholders.

In some practices, not all partners gave a convinced adherence to the projects and, when those projects were selected, they refused to cooperate in the projects' realization (see, e.g., Three Parks Chain). In other practices, where stakeholders were belatedly involved, they abandoned the project or even opposed it (see, e.g., Not only Lake). In more than half of the practices, which coincide with most of those aimed at building operators' networks and partnerships for the local economic systems innovation, there have been difficulties in relationship, decision sharing, and activity management between leader and economic operators (see, e.g., Participative Guarantee). Finally, there are practices (Resilient Gardens, The Door of the Park, and Crazy-lienza) that have been conditioned by difficulties in involving local communities and in triggering mechanisms of appropriation and taking charge of the places and resources revitalized by the project.

The carried out actions have made it possible to achieve almost completely the *objectives* indicated by the practices, while they have had limited effects in achieving the stated *aims* of contrasting the higher scale phenomena of risk, land abandonment, economic and environmental impoverishment (see Table 6.2), as a consequence of the gap between *aims* and *objectives* indicated earlier.

The practices that indicated as *objective* the innovation of networks/systems/tools for sustainable resource management (Eco-Food Chain and GE.COO.FOR) and of local economies (Three Parks Chain, Participative Guarantee, The Blond Goat, Not only Lake) promoted preparatory actions and preliminary experiments, the improvement of the integration and coordination among stakeholders, the launch of training and raising awareness activities, and the use of new tools, procedures, and protocols, but they were not able to reach the full implementation of such actions, with reduced effects on the contrast of risk, abandonment, and economic and environmental impoverishment phenomena for which they were thought. The other practices, oriented to the environmental recovery and enhancement and to the reappropriation and shared management of degraded areas (Resilient Gardens, The Door of the Park, Crazy-lienza), contributed to the improvement of environmental conditions with physical restoration and recovery interventions of specific areas and to the sustainable fruition promotion, with difficulties in triggering stable and lasting management and with limited effects on the contrast of the abandonment phenomena for which they were thought.

Table 6.2 Main aims and objectives envisaged and achieved by resilience practices.

Resilience practices	Aims	Envisaged objectives	Achieved objectives
Eco-Food Chain GE. COO. FOR.	Contrast to environmental damages	Activation of sustainable resource management system	Promotion of sustainable resource management systems tool
Three Parks Chain	Contrast to economic and environmental impoverishment	Activation of innovative local economic system	Raising awareness toward activation of innovative local economic system
Participative Guarantee	Contrast to economic and environmental impoverishment	Activation of innovative tools to support the food supply chain	Experimentation of innovative tools to support the food supply chain
The Blond Goat	Contrast to abandonment and economic and environmental impoverishment		
Not only Lake	Contrast to abandonment and economic and environmental impoverishment	Activation of innovative local economic system	Raising awareness toward activation of innovative local economic system
Resilient Gardens The Door of the Park Crazy-lienza	Contrast to abandonment	Recovery, reappropriation, and shared management of degraded contexts	Environmental recovery of specific sites and sustainable fruition promotion

6.3.3 The assessment of the overall impact of resilience practices on reference systems

The effects' assessment of the practices on the *reference systems* resilience capabilities is referred to the achievement of their *goals* and *objectives* and not to the general resilience capacity of the *reference system*. This is to prevent the rating squashing on low levels by making the evaluation not useful to understand the impact of these practices, considering that those practices were not designed to solve the critical issues of the *reference system*, but to make a contribution in solving those critical points.

The *secondary resilience assessment factor* on which the practices have most affected is *information/knowledge*, followed by *awareness/public spirit/self-organization* (see Fig. 6.2). This outcome is probably due to the subjects admitted to the call *Resilient Communities*, which in most cases have been nonprofit organizations whose field of action concerns raising awareness, involvement, and capacity building of communities and stakeholders. On the other hand, these factors refer to the initial phases actions of a community activation process that have been considered preparatory to the project realization.

The less incident *secondary assessment factor* is *redundancy*, which has been considered in a single practice. This is probably because the functions and resources oversizing in a territorial system come into conflict with the efficiency in the resources use. Thus *redundancy* tends to be taken as an *objective* when there is a need to reduce the territorial system vulnerability with respect to the stress or shock risk. In fact, the increase in the *redundancy* of a territorial system can be useful to guarantee the energy supply stability, to maintain the road connections in marginal areas exposed to seismic phenomena, or to achieve a good stability of the banking system in the event of economic crisis. Furthermore, an increase in the territorial system resilience capacity can lead to an increase in the *redundancy* in the higher scale system. For example, the efficiency increase

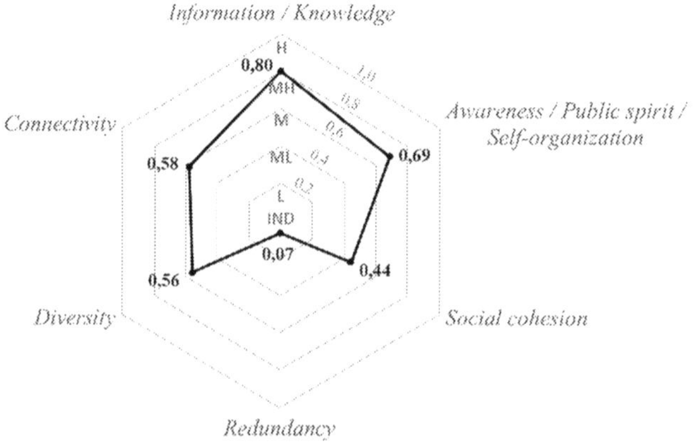

Figure 6.2 Average ratings attributed to secondary resilience assessment factors for the set of practices.

in an area in the waste treatment can lead to an increase in the *redundancy* of the territorial system in which that area is inserted, because it can support other areas belonging to the system in the case of critical situations due to plant failures or unexpected increases in waste production.

The monitored practices did not have to face scarcity and vulnerability of resources problems such as to jeopardize the functioning of the *reference systems* in which they intervened, so they aimed above all to increase and improve the existing resources endowment.

To understand how the set of practices affected the *secondary resilience assessment factors*, considered both as a whole and individually, the frequency distribution of the five rating classes attributed to the set of practices with respect to resilience factors was analyzed. This distribution assumes a trend represented by a Gaussian curve slightly shifted toward the higher classes (see Fig. 6.3), which indicates that the analyzed practices have had a significant impact on *secondary factors*.

In particular (see Table 6.3), the practices' performance was more frequently based on the average rating, due to a high concentration on *connectivity* and *diversity* factors, and with a significant recurrence on the MH level, mainly due to *information/knowledge* and *awareness/public spirit/self-organization* factors. The *information/knowledge* factor also presents three cases in which it has a high performance; excluding *redundancy*, the factor on which the practices have had a low influence is *social cohesion*.

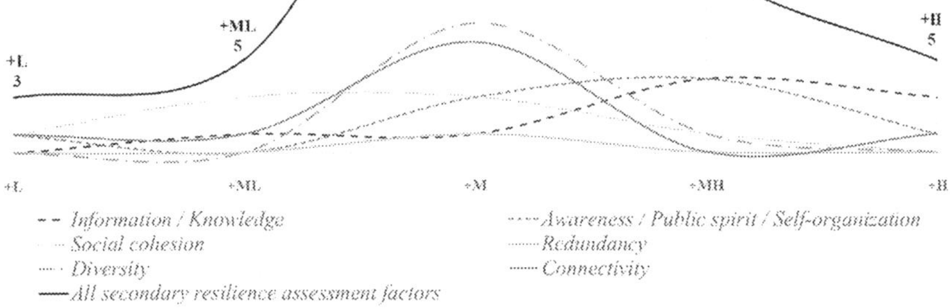

Figure 6.3 Trends in the frequency of ratings attributed to secondary resilience assessment factors.

Table 6.3 Frequencies of rating attributed to primary and secondary resilience assessment factors.

Resilience assessment factors	Rating classes					
	IND	L	ML	M	MH	H
Information/Knowledge	0	0	1	1	4	3
Awareness/Public spirit/Self-organization	0	1	0	3	4	1
Social cohesion	1	1	3	3	1	0
Redundancy	8	0	0	1	0	0
Diversity	1	0	0	7	1	0
Connectivity	0	1	1	6	0	1
Secondary resilience assessment factors	10	3	5	21	10	5
Robustness	0	0	3	3	3	0
Adaptability	0	0	2	3	4	0
Primary resilience assessment factors	0	0	5	6	7	0
Resilience	0	1	2	3	1	2

H, High; *IND*, indifferent; *L*, low; M, medium; *ML*, medium-low; *MH*, medium-high.

Therefore the set of practices leads to an increase and an improvement of *information/knowledge*, *awareness/public spirit/self-organization*, *connectivity*, and *diversity* factors. This is because they have provided for the collection and systematization of information on the areas of intervention, relating both to the actual and expected territorial, socioeconomic and environmental characteristics and dynamics and to the stakeholders. This cognitive step has favored the identification of existing and potential phenomena and critical factors, and the verification and integration of the critical issues indicated in the project proposals. Furthermore, this information has allowed to assess the practices effects and the stakeholders and communities' behaviors on the critical elements and to verify the priority and feasibility of the contrast actions envisaged in the planning phase.

Finally, the practices contributed to identifying and developing the useful skills and knowledge (management, coordination, planning, teaching, intervention, problem-solving, etc.) to tackle the critical issues.

On the basis of the acquired information and knowledge, the set of practices has aimed at spreading knowledge and improving the understanding of the critical elements to contrast, of the strategies to achieve the envisaged objectives, and of the effects, positive and negative, that individual behaviors and the collective actions can generate. The practices also favored the increase of civic and economic communities' awareness on the potential contribution of individual behaviors, even if they had difficulties in triggering mechanisms of self-organization and proactive reactions of citizens and communities for the concrete realization of actions.

The practices have promoted the meeting, the exchange, and the actions sharing between social and economic actors, coming to organize, in some cases, networks of economic operators functional to the producer—consumer chains activation and, in other cases, extensive and articulated relationship networks between different actors (environmental and others kind of associations, schools, agricultural producers, scholars, fragile subjects, religious organizations, social cooperatives, municipal technicians, park authorities, etc.) interested in sharing, with variable geometries and different times and ways, specific actions of larger projects.

Most of the practices improved the endowment and the performance of the *reference systems* resources to respond to the existing critical issues through the involvement of actors and skills (planning, management, trading, etc.) and the use of new tools.

The improvements in *secondary resilience assessment factors* and the effects of their interactions have led to an increase in *adaptability* and *robustness* in *reference systems*, with significant positive effects on *resilience*. In particular, the set of practices has contributed to a significant improvement in the *adaptability* of the elements involved, with a higher recurrence on the medium-high level (4) respect to medium (3) and medium-low (2) rating classes, and, to a slightly lesser extent, to a strengthening of their *robustness*, with a homogeneous recurrence (3) on these three rating classes (see Table 3.3).

The *information/knowledge* and *awareness/public spirit/self-organization* factors have significantly influenced the improvement of *adaptability*, the very positive performances of which indicate that the practices, to achieve their goals, have significantly increased knowledge, skills, and the ability to organize and activate communities and stakeholders, also by experimenting with innovative solutions and taking into account the conditions of uncertainty regarding the evolution of *reference systems*. The *information/knowledge*, *diversity*, and *connectivity* factors contributed to the strengthening of *robustness*, where practices have increased, through structural and nonstructural interventions, the availability, stability, and reliability of resources useful for dealing with stress and shock situations (see Table 6.4).

The positive repercussions on *robustness* are more related to the effects of actions on *secondary factors*, such as recovery and reactivation actions of

Table 6.4 Ratings of the primary and secondary resilience assessment factors and of the resilience of the assessed practices.

Resilience practices	Secondary factors						Primary factors		*Resilience*
	Information/ Knowledge	Awareness/ Public spirit/ Self-organization	Social cohesion	Redundancy	Diversity	Connectivity	Robustness	Adaptability	
Three Parks Chain	MH	MH	ML	–	M	M	MH	MH	H
GE.COO. FOR.	MH	MH	–	M	M	M	MH	MH	H
Crazy-lienza	H	H	M	–	M	M	M	M	MH
Resilient Gardens	MH	M	MH	–	MH	H	ML	MH	M
Not only Lake	H	M	ML	–	M	M	M	MH	M
The Blond Goat	H	M	M	–	M	M	MH	M	M
Participative Guarantee	M	MH	M	–	M	ML	M	ML	ML
Eco-Food Chain	MH	MH	ML	–	–	L	ML	M	ML
The Door of the Park	ML	L	L	–	M	M	ML	ML	L

H, High; *IND*, indifferent; *L*, low; M, medium; *ML*, medium-low; *MH*, medium-high.

the social, organizational, and economic resources, while the strengthening of *adaptability* is more linked to the materialization of "emergent properties," such as the increase in the ability to manage uncertainty and to respond with changes to the emerged needs and the improvement of institutions in the ability to share innovative solutions and of the involved actors in making up for the institutions' shortcomings with subsidiary activities.

The improvements in *adaptability* and *robustness* and the effects of their interactions generate positive repercussions on the *resilience* of *reference systems*, with ratings above the medium level in most practices (6), two of which with high ratings (see Table 6.3). The performance of *resilience* is higher than the ratings attributed to *primary resilience assessment factors* and highlights the synergies that the interactions between *adaptability* and *robustness* can achieve (see Table 6.4).

6.3.4 The assessment of the impact of single resilience practices on related reference systems

The effect's evaluation of the single practices on the related *reference systems* resilience capacity, concerning the achievement of the *objectives* of the practices themselves, shows a very differentiated rating, which varies from maximum to minimum classes of the *standard rating scale* (see Table 6.4).

The Three Parks Chain and GE.COO.FOR practices have significantly acted on most *secondary resilience assessment factors*, with the exception of *redundancy* or *social cohesion* and with greater intensity on *information/knowledge* and *awareness/public spirit/self-organization*. This thanks to the introduction of some operational tools (online platforms, protocols and procedures, information, assistance desks, etc.) that are particularly effective in achieving the *objectives* of the practices. Thus the effects on *secondary factors* have significantly affected both the *primary factors* and *resilience*.

The Crazy-lienza practice had an impact similar to the previous practices due to its ability to modify *secondary resilience assessment factors*, but, having acted in a very restricted and isolated territorial context, it had more contained effects both on *primary factors* and on *resilience*.

Also the Resilient Gardens, Not only Lake, and The Blond Goat practices affect almost all *secondary factors*, but with less effectiveness on *awareness/public spirit/self-organization*. These practices had very robust leaders who in some cases made up for the difficulties in stakeholders actively engaging, while in other cases, they faced difficulties with good adaptive ability, activating new corrective solutions.

The crucial role of the leader partner has characterized also the practice The Door of the Park, the implementation of which has encountered many difficulties due to the conflicting relationship between the leader himself, on the one hand, and the project partners and the affected community, on the other hand.

Finally, the Participative Guarantee and Eco-Food Chain practices, which had similar effects on *secondary factors* and on *resilience* and inverted effects on the two *primary factors*, concentrated their action on the analysis, design, and transmission of knowledge within the management of a certification process of an organic farming activity and of a sustainable collection system of game meat. However, these practices have been inadequate in the construction of the stakeholders and operators network involved in experimenting with the agricultural certification and the supply chain for the consumption of game meat, reducing their positive effects on *secondary factors* and *resilience.*

From the results of the assessment, three profiles of "impulse" to the resilience were outlined, to which the monitored practices were placed. All three profiles are united by the propensity for change and coevolution and distinguished with respect to practice effectiveness in generating a resilient change on the related *reference systems.*

The first profile is called "adequate impulse to change" and includes the practices that succeed in triggering a solid and lasting change in the *reference system.* They are practices that affect most of the *resilience assessment factors* concerning the critical issues they intended to solve. Those practices have built a good level of sharing of objectives and actions between project partners and stakeholders and have created the conditions to ensure the continuity of the project implementation even in the event of changes in the partnership. This profile includes the Three Park Chain, GE.COO. FOR, and Crazy-lienza practices, although the latter has a lower effectiveness in the change because it underestimates the economic sustainability of its action, making its duration depend exclusively on voluntary ethical/cultural adherence.

The second profile is called "subordinate impulse to change" and includes the practices that make the achievement of a solid and lasting change of the *reference system* partly dependent on external conditions related to *resilience factors* on which it was not possible to act. This profile includes those practices—such as the Resilient Gardens, Not only Lake, The Blond Goat, and The Door of the Park—that aimed to improve and innovate the traditional tools and methods of intervention without sufficiently affecting on awareness and project sharing, finding difficulties to actively involve partners, stakeholders, and the community.

The third profile is called "embryonic impulse to change" and includes the practices that hypothesize a possible change starting from an idea, often already tested in other contexts. This profile includes the practices—such as the Participative Guarantee and Eco-Food Chain—that deepened the knowledge of the territorial system, acted on the awareness of the stakeholders, tried to improve the available resources, favored the implementation of innovative processes, without, however, succeeding to implement interventions able to obtain significant changes in resilience in *reference systems*.

References

Carpenter, S., Walker, B., Anderies, J. M., & Abel, N. (2001). From metaphor to measurement: Resilience of what to what? *Ecosystems*, *4*(8), 765–781.

Chelleri, L. (2015). Urban resilience trade-offs: Sfide derivate dall'applicazione di un approccio integrato alla resilienza urbana. GSSI cities Working Papers. *SSRN Electronic Journal*, *20*, 1–25. Available from https://www.researchgate.net/publication/275946726_Urban_Resilience_Trade-Offs_Sfide_Derivate_Dall'Applicazione_Di_Un_Approccio_Integrato_Alla_Resilienza_Urbana.

Chuvarajan, A., Martel, I., & Peterson, C. (2006). *A strategic approach for sustainability and resilience planning within municipalities. Thesis submitted for completion of Master of Strategic Leadership towards Sustainability*. Karlskrona, Sweden: Blekinge Institute of Technology.

Cutter, S. (2016). The landscape of disaster resilience in the USA. *Natural Hazards*, *80*(2), 741–758.

Davoudi, S., Brooks, E., & Mehmood, A. (2013). Evolutionary resilience and strategies for climate adaptation. *Planning Practice and Research*, *28*. Available from https://doi.org/10.1080/02697459.2013.787695.

Davoudi, S., Shaw, K., Haider, L., et al. (2012). Resilience: A bridging concept or a dead end? "Reframing" resilience: Challenges for planning theory and practice interacting traps: resilience assessment of a pasture management system in Northern Afghanistan Urban Resilience: What does it mean in planning practice? Resilience as a useful concept for climate change adaptation? The politics of resilience for planning: A cautionary note. *Planning Theory and Practice*, *13*(2), 299–333. Available from https://doi.org/10.1080/14649357.2012.677124.

Fleming, J., & Ledogar, R. J. (2008). Resilience, an evolving concept: A review of literature relevant to Aboriginal research. *Pimatisiwin*, *6*(2), 7.

Folke, C., Carpenter, S., Walker, B., Scheffer, M., Chapin, T., & Rockström, J. (2010). Resilience thinking: Integrating resilience, adaptability and transformability. *Ecology and Society*, *15*(4).

Fusco Girard, L. (2011). *Multidimensional evaluation processes to manage creative, resilient and sustainable city*. Aestimum, Firenze: Firenze University Press, ISSN:1592–6117.

Holling, C. S. (1973). Resilience and stability of ecological systems. *Annual Review of Ecology and Systematics*, *4*, 1–23.

Longstaff, P., Armstrong, N., Perrin, K., Parker, W., & Hidek, M. (2010). Building resilient communities: A preliminary framework for assessment. *Homeland Security Affairs*, *VI*(3).

MacKinnon, D., & Derickson, K. D. (2013). From resilience to resourcefulness: A critique of resilience policy and activism. *Progress in Human Geography, 37*(2), 253–270.

Marcus, L., & Colding, J. (2014). Toward an integrated theory of spatial morphology and resilient urban systems. *Ecology and Society, 19*(4), 55.

Martin, R., & Sunley, P. (2015). On the notion of regional economic resilience: Conceptualization and explanation. *Journal of Economic Geography, 15*, 1–42, United Kingdom: Oxford Academic.

Olazabal, M., Chelleri, L., & Sharifi, A. (2018). Is connectivity a desirable property in urban resilience assessments? In Y. Yamagata, & A. Sharifi (Eds.), Resilience-oriented urban planning. Theoretical and empirical insights, *lecture notes in energy* ((65). Switzerland: Springer.

Olsson, L., Jerneck, A., Thoren, H., Persson, J., & O'Byrne, D. (2015). Why resilience is unappealing to social science: Theoretical and empirical investigations of the scientific use of resilience. *Science Advances, 1*(4).

Pendall, R., Foster, K. A., & Cowell, M. (2010). Resilience and regions: Building understanding of the metaphor. *Cambridge Journal of Regions, Economy and Society, 3*, 71–84.

Portugali, J. (2000). *Self-organization and the city*. Berlin: Springer.

Reghezza-Zitt, M., Rufat, S., Djament-Tran, G., Le Blanc, A., & Lhomme, S. (2012). What resilience is not: Uses and abuses. *European Journal of Geography*, Cybergeo: UMR 8504 Géographie-cités.

Sharifi, A., & Yamagata, Y. (2018). Resilient urban form: A conceptual framework. In Y. Yamagata, & A. Sharifi (Eds.), Resilience-oriented urban planning. Theoretical and empirical insights, *lecture notes in energy* (65). Switzerland: Springer.

Sharifi, A., Chelleri, L., Fox-Lent, C., Grafakos, S., Pathak, M., Olazabal, M., & Yamagata, Y. (2017). Conceptualizing dimensions and characteristics of urban resilience: insights from a co-design process. *Sustainability, 9*(6), 1032.

Walker, B., & Salt, D. (2012). *Resilience practice. Building capacity to absorb disturbance and maintain function*. Island Press.

CHAPTER 7

The impacts of resilience practices on local food systems: evidences from Italian case studies

Giampiero Mazzocchi[1], Catherine Dezio[2] and Davide Marino[3]
[1]Department of Policies and Bio-economics, Council for Agricultural Research and the Analysis of Agricultural Economics, Rome, Italy
[2]Department of Architecture and Urban Studies, Politecnico di Milano, Milan, Italy
[3]Department of Biosciences and Territory, University of Molise, Pesche, Italy

7.1 Introduction

Systems where social, economic, ecological, cultural, political, technological, and other components are strongly linked are known as socioecological systems (SESs), emphasizing the integrated concept of the "humans-in-nature" perspective. SESs are truly interconnected and coevolving across spatial and temporal scales, where the ecological component provides essential services to society such as supply of food, fiber, energy, and drinking water. As a result, SESs have become an emerging focus in scientific and policy arenas (Petrosillo et al., 2015).

The concept of resilience in ecological systems was first introduced by the Canadian ecologist C.S. Holling (Holling, 1973) to describe the persistence of natural systems in the face of changes in ecosystem variables due to natural or anthropogenic causes. Resilience has been defined in two ways in ecological literature:

- As the time required for an ecosystem to return to an equilibrium or steady state following a perturbation (which is also defined as stability by some authors). This definition of resilience is used in other fields such as physics and engineering, and hence has been termed "engineering resilience" by Holling.
- As "the capacity of a system to absorb disturbance and reorganize while undergoing change so as to still retain essentially the same function, structure, identity, and feedbacks" (Walker et al., 2004).

[ECO]systems of Resilience Practices
DOI: https://doi.org/10.1016/B978-0-12-819198-9.00014-4

Currently, the relationship among agriculture, food, and cities is experiencing profound transformations that led us to reflect on the causes, effects, factors, and processes (European Commission, 2017; Marino et al., 2014; Marino et al., 2018; MEA, 2005). It is clear that patterns of food production, consumption, and well-being develop not only from economic and social relations of urban systems but also depend on their capacity to sustain them (i.e., their resilience) (FAO, 2017; Steffen et al., 2015). In this chapter the authors keep together the two aforementioned dimensions of resilience (time and effectiveness to restore the function of a system after a perturbation). They do it analyzing the capacity of 50 practices related to food–city relationships, belonging to four dimensions (food system, territorial system, social system, risk management), to adapt themselves to the changing endogenous and exogenous factors impacting their activities and objectives, as well as the strategies adopted to overcome obstacles and barriers encountered during the implementation. The research represents an implementation of the impact evaluation framework to measure urban resilience in food practices (Dezio & Marino, 2018). In that context the research adopted the coevolutive approach, which recognizes territorial dynamics as products of biunivocal relations between social and environmental components. Particularly, the authors chose to adopt the coevolutionary paradigm of Norgaard (Kallis & Norgaard, 2010; Norgaard, 1984), which states that economic history is a process of adaptation to changes in the environment, and that such transformations are bijective. For this dynamic relationship between environmental and social systems, the coevolutionary paradigm is the key to better understanding the transformations of SESs, such as agro-food systems. Several SES frameworks can be applied to the analysis of food systems, objective of the present chapter. They reflect the variety of research fields that can be applied to conceptualize the SESs. Among them, the driver–pressure–state–impact–response (DPSIR) conceptual model is a framework that aims at depicting the cause–effect relationships between environmental and human systems and at assessing the social and ecological problems of systems subject to anthropogenic influence through the DPSIR linkages. Foran et al. (2014) provide a synthesis of what is intended with "SES". They state that SES visualizes the human–environment interface as a coupled system in which socioeconomic and biophysical driving forces interact to influence food system (and subsystem) activities and outcomes, both of which subsequently influence the driving forces (p. 90). Thus complexity and dynamicity are considered inherently features of the SES that continuously adapt to response to internal or external pressures. This understanding of food systems as SESs helps answering questions about their sustainability. In this context, when referring to sustainability of food systems, all aspects of the food system are considered. These include urban and peri-urban agriculture; strengthening the rural–urban interface to ensure connections between rural supplies and urban contexts; taking into account street food, retailers, food processing and distribution, nutrition, and health; and linkages to water, waste, transport, and energy systems. Those food systems' activities and properties emanate, in turn, in three main sets of outcomes such as food and nutrition security (availability, access, utilization), social welfare, and environmental capital (Prosperi et al., 2019).

Furthermore, various elements of food systems are altered by, and actively impact, the socioeconomic and environmental conditions of the system across local, regional, and global levels. Food security depends on multiple biophysical drivers—such as land and water resources degradation, biodiversity loss, sea-level rising—and economic and social stresses—such as demographic dynamics, technological innovation, economic trends, and social changes—which interact with each other and then impact, individually or concurrently, different aspects of the food system. Such socioeconomic and biophysical drivers might impact food security directly or indirectly, positively or negatively, and change over time; when a food system fails to deliver people food security, considered its primary outcome, the system can be considered vulnerable (Prosperi et al., 2019). Therefore food systems can be vulnerable, and resilient, to a set of stressors (Adger, 2006) such as environmental pressures, socioeconomic instabilities, and institutional and policy factors (Turner et al., 2003).

7.2 Agro-environmental practices for resilience

The Observatory of Resilience Practices (ORP) consists of several axes. The Mapping Path, which is the mapping of collected practices and is a strategic axis of the ORP, requires a tight integration between the construction of a conceptual framework, the direct involvement of the promoters of initiatives, and an innovation in the identification of methodological and design tools. The Mapping Path provided the opportunity to the whole local community to propose and subsequently validate practices that fall within Holling's definition of ecosystem resilience (Holling, 1973). "Ecosystem resilience" assumes a dynamic and evolutionary vision of possible response mechanisms of complex territorial systems (Holling, 1973).

Strengthening resilience properties of complex systems (such as territorial ones) allows the same systems to activate responses, adaptation, and change in face of exogenous or endogenous disturbances. With strengthened resilience properties, systems may reconfigure toward new equilibrium conditions, as well as acquire new capacities to face the future.

With that theoretical reference point the Observatory collected 100 practices: 24 at Lombardy scale, more than 40 from within national context (i.e., Italian scale), and 30 scientific applied research studies. The practices are concerned with varying themes, dealing with both social and environmental dynamics, but they have the same objective of strengthening resilience against fragilities recognized in the system as follows: the issue of social inclusion, the support for disadvantaged groups, the prevention of hydrogeological instability, the support for the local economy, the redevelopment of degraded public spaces, the protection of endangered species, and much more.

Referring to the aforementioned research by Dezio and Marino (2018), the authors decided to filter 50 out of the 100 practices collected concerning food and agro-ecosystems because, with a starting point of Holling's definition of ecosystem resilience, agro-ecosystems turn out to be integrated SESs suitable to be considered bearers of a key role for the whole territorial system. Therefore this sample can become an investigative opportunity on the evaluation method of the effectiveness of bottom-up practices. The 50 practices that Dezio and Marino investigated use food and agriculture to strengthen resilience of different fragilities. The practices include strategies for the inclusion of urban agriculture for the reappropriation of degraded territories and/or for the support of disadvantaged groups, the launching of alternative quality practices to traditional production—distribution systems establishment of support systems for local quality agriculture and networks, strategies to restore traditional crops, and many others.

7.3 Materials and methods

Following a detailed analysis of the individual actions conceived for each practice, Dezio and Marino identified four macro-categories, representing areas of resilience on which the practices are oriented. The four categories also coincide with different reference systems:

- food system, that is, the practices that aim to support local short supply chains;

- territorial system, that is, the practices aimed at redevelopment, enhancement, and protection of urban space, territory, and landscape;
- social system, which includes practices that aim to raise awareness and strengthen the resilience of the local community; and
- risk management, a category that focuses on the broad concept of "risk" (e.g., hydrogeological risk, risk of loss of animal species, consumption of soil, risk of economic crisis).

It has to be remarked that all the selected practices refer to agro-environmental systems, yet in their application they impact directly or indirectly on one or more of the four identified systems.

Following the premises of the research, that is, assess the capacity of resilience practices to achieve the expected impacts and to overcome barriers and obstacles, the authors contacted the focal points of the 50 practices related to agro-environmental practices to hand them a questionnaire aimed at investigating the mentioned aspects. The research has been conducted through a computer-assisted web interviewing survey, and the questions have been prepared so as to guarantee a comparison between the answers (through Likert scales) and a certain degree of freedom (through open-ended answers and keywords).

7.4 Results and discussion

Overall, the rate of responses has been the 34%, that is, 17 practices have been surveyed. On average, the 70.6% of the projects have been concluded at the moment of the survey (i.e., activities have been concluded and results and impacts have been monitored and assessed). The 23.5% of the practices have not yet been concluded because the activities envisage a mid-long duration, while only one practice is still in an intermediate phase of realization. The large part of the sampled practices (88.2%) has been financed by a private banking foundation, while the remaining have received fundings from both private and public funds or have not perceived any grants. In the large majority of the cases (82.4%), the projects have been concluded within the first round of funding, while the 17.6% of the practices have requested and obtained a second round of financing for their activities.

Overall, the self-assessment reflects a discrete effectiveness of the projects in terms of achievement of the prefixed objectives and expected impacts that respectively show averages of 3.82 and 3.65 on a Likert scale from 1 to 5 (1 = very low; 5 = very high). Nevertheless, it has to be remarked that in only two cases the practices' representants have answered

"Very high" to the two answers on achievement of objectives and expected impacts, while, on the other hand, one respondents has reported "Low" for the two answers. On average, the impacts of the activities carried out within the project have solved the critical factors that led to the designing of the project to a medium extent (3.29 on a Likert scale from 1 to 5), with a deviation standard of 0.85. The latter is even higher relatively to the question that seeks to indagate whether and to what extent unexpected factors that took place during the realization of the project have been faced and solved. Indeed the standard deviation is 1.25 and the average mark is 3.00 that represent a medium result. Similarly, according to respondents, obstacles met during the realization of the activities have been solved to a medium extent (3.12), reflecting the fact that more adaptability and flexibility should have been occurred to face unexpected events. Indeed, respondents answered that obstacles met during the activities' implementation have produced a negative impact on the project effectiveness to a medium-low extent (2.59), showing that practices could have been more adaptive. Nevertheless, higher marks (3.53) have been reported as regards the contribution of the practice to the strengthening of the resilience in the reference system.

According to the theoretical background depicted in the introductive section of the chapter, a qualitative analysis of relevant factors emerged from practices is highlighted here. In particular, have been surveyed and critically analyzed the following elements: criticalities that led to the design of the projects; successful elements occurred during the implementation of the practice; obstacles encountered; solutions found; unexpected positive and negative impacts.

7.4.1 Criticalities

According to the results deriving from the survey, it is possible to depict some relevant fragility factors, the motivations, and the elements that led to the conception and development of the practice. Basically, in this section are presented the initial pressures that, applying on the social-ecological local systems, were acting a significative modification of the environmental and social balances. Four different clusters of risks and criticalities have been identified: environmental risks and resources competition; territorial balances; economic and market hazards; loss of social capital. In Table 7.1 have been reported the specific issues identified for each cluster of risks and threats.

Table 7.1 Risks, threats, and criticalities that have led to the design of the practices.

Categories	Specific issues
Environmental risks and resources competition	Anthropization of natural areas Competition between wildlife presence and human activities Conflicts and damages Hydrogeological instability Poor usability of the territory Soil consumption Hydraulic risk Lack of intermunicipal responses to environmental risks Protection of agricultural areas with high naturalistic and environmental value soil contamination Air and noise pollution Loss of soil fertility and biodiversity Dissemination of alternative techniques to widespread conventional practices building risk on public areas of considerable value Climate change and impacts on the environment Conflicts between livestock farmers (sheep and goats) and large predators (bear and wolf)
Territorial balances	Rapid social changes on an urban scale Mountain areas abandonment and lack of care for the territory Socioeconomic isolation of marginal areas Risk of abandonment of urban green areas
Economic and market hazards	Farms shutdown Lack of short and sustainable supply chains Agricultural production in decline Abandonment of terraced areas and related economic recovery Poor incentives to work in agriculture Difficulties in creating a commercial network Availability of some products only in alternative supply chains.
Loss of social capital	Individualism, isolation, and mistrust Insufficient involvement of noninstitutional actors in climate change adaptation measures Loss of social cohesion Poor sensitivity of the community and institutions on local supply chains Need to acquire agronomic skills

Credit: authors' elaborations.

7.4.2 Successful elements

During the implementation of practices, some factors, resources, conditions, and actions contributed to the success of the project according to the set objectives (Table 7.2).

Table 7.2 Factors, resources, conditions, and actions contributed to the success of the project according to the set objectives.

Categories	Specific issues
Project Management	Determination and cohesion of the working groups Strong motivation and participation of the partners Constant coordination Technical simplicity of realization Low construction costs Participatory planning Focus on a single supply chain
Interinstitutional management	Intermunicipal and multistakeholder approach Increased risk response capacity by civil protection and municipalities Involvement of a large territorial network (institutions and civil society) Availability of the municipalities involved on the education/education front
Community involvement	Involvement of local communities Bottom-up structuring proposals for change Increase in social capital and relations between different ethnic communities Greater empowerment Involvement of schools Birth of a virtuous social system together with the economic one Peaceful and constructive meeting between city and rural world Interest from citizens
Recovery of socioecological systems	Reactivation of ecological functions of degraded and abandoned areas Creation of local and innovative food chains Promotion of food and wine tourism Recovery of the ecological function of degraded and/or abandoned areas Reuse and recycling of natural resources

Credit: authors' elaborations.

7.4.3 Obstacles

In the survey the referents have been asked to describe the unforeseen events that, during the implementation of the practice, have changed the ecological, economic, or social conditions that motivated the conception and development of the practice. These unexpected obstacles are summarized in Table 7.3.

Table 7.3 Unforeseen events that, during the implementation of the practice, have changed the ecological, economic, or social conditions that motivated the conception and development of the practice.

Categories	Specific issues
Balances and political relationships	Political transitions and change of political representatives
	Conflicts between citizens residing in hydraulic risk areas and municipal administrations
	Relationship difficulties with local authorities
Participation	Distrust between ethnic communities
	Poor cooperation from farmers and breeders
	Difficulty interacting, focusing on one's goals
	Poor response from citizens to initiatives organized on the ground
	High average age of landowners
	Excessive individualism of some beneficiaries
	Conflict between beneficiaries (gardeners)
	Difficulty in making participatory and democratic management speak reality with more hierarchically organized partners
	Low participation and involvement of the local community
Distrust and path dependency	Understanding of the project potential and benefits at the local economic level
	Training of people reluctant to change their usual behavior
	Bureaucratic and administrative resistance to changes in the local system
	Limited size and organizational capacity of farms
	Network project development is perceived as a greater investment of managerial and organizational resources and an economic risk
	Cultural and technical preconceptions by professionals
	Distrust of nonindustrial or manufacturing development models

(Continued)

Table 7.3 (Continued)

Categories	Specific issues
Endogenous and exogenous project difficulties	Scarcity of economic and human resources Redesign of the intervention due to unfavorable environmental/ecological conditions Participatory planning has lengthened the time and led to numerous deviations from the initial project Climate changes that make crop patterns unpredictable Difficulty working with volunteer people with various needs Longer use of time than budgeted Pests and adverse climatic conditions that have reduced the quantities produced and the possibility of conveying products outside the GAS Land made available inadequate and unattractive (too small and close to industrial area) Initial availability of making available of land subsequently adjusted due to the absence of concrete possibilities

GAS, Gruppo di Acquisto Solidale, that is Solidarity Buying Group.
Credit: authors' elaborations.

7.4.4 Solutions

To properly respond to the new challenges occurred during the implementation of the project due to the unexpected obstacles and events, the practices have identified some solutions to reestablish a new socioecological balance (summarized in Table 7.4).

7.4.5 Unexpected impacts

When analyzing practices according to the socioecological theoretical frameworks, unexpected impacts are considered elements that alter the socioeconomic and environmental conditions of the system across local and regional levels. Such socioeconomic and biophysical elements might impact surveyed food systems directly or indirectly, positively or negatively, and change over time. For this matter the surveyed representants have been asked to highlight possible unexpected impacts produced by the implementation of the practice, or impacts that were expected to a

Table 7.4 Solutions found to reestablish a new socioecological balance after the occurrence of unexpected events and obstacles.

Categories	Specific issues
Project training	Training meetings Production of specific and addressed materials Transparency of relationships Widespread information Clear identification of the project representatives Constancy of commitment Communication plan Increase in the number of workers
Mediation and flexibility	Adaptation and flexibility of the project proposal Listening to the needs of the stakeholders and willingness to review the methods of intervention Adaptation of activities to individual contexts in response to specific critical issues Rebalancing of financial balances Mediation skills Conflict facilitation
Multistakeholder interactions	Work on actions that make subjects interact Establishment of associations
Events and stimulation of participation	Gastronomic initiatives for the presentation of local products Organization of initiatives in direct contact with citizens Creation of practical examples Concreteness of actions

Credit: authors' elaborations.

lesser extent compared to the ex post results. The analysis of the answers has been summarized in Table 7.5.

7.4.6 Discussion

The surveyed practices have highlighted some relevant elements that compose a complete framework for analyzing the flow of drivers, obstacles, solutions, impacts, and successful elements related to the implementation of the resilience practices. This stream has been represented in Fig. 7.1. In this figure, all the elements objects of investigation have been put in relation each other and placed within a framework where SESs, and their related feedbacks, continuously and constantly act on the

Table 7.5 Unexpected impacts produced by the implementation of the practice.

Categories	Specific issues
Knowledge increase	Strengthening of participatory planning capacity between Municipalities
	Increased awareness of technicians and municipal administrators on specific issues (hydraulic risk)
	Creation of specific skills
	Introduction of principles and practices of agroforestry/ syntropic agriculture, especially associated horticulture
	New environmental awareness of younger generations
Self-standing continuation	Support for the development of a local resale network
	Project autonomy of some citizens or stakeholders
	Continuation of events and parties aimed at citizens
	Approaching the project by similar experiences
Supralocal spread	Dissemination of the project at the supralocal and national level
	Strengthening an intermunicipal network of civil protection groups for better management of environmental risks
	Interest from neighboring municipalities that led to nominate a similar project, but which affects larger territories
	Networking and cooperation with other similar experiences
Project activation	Activation of other funds to follow up on the activities started
	Community interest in lending itself to maintenance
	Involvement of migrants within the project
	Activation of social and solidarity economies
	Birth of other projects that share the same spirit of the initial project
	Great media attention
	Participation of volunteers and beneficiaries (shepherds)

Credit: authors' elaborations.

implementation of the practices, modifying balances, putting new unforeseen pressures, forcing the projects to adapt to the new equilibrium. Fig. 7.1 also summarizes the main categories of the survey, thus combining the theoretical and methodological models of interactions between SESs and food systems and the results of the investigation in a single graph.

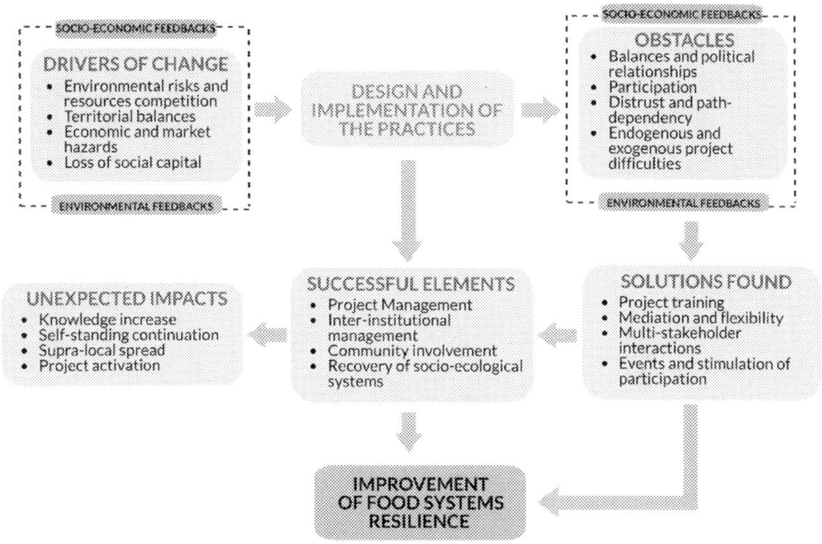

Figure 7.1 Interactions between implementation of resilience practices, their impacts, and food systems. *Authors elaborations.*

The fragile factors, the motivations, and the elements that led to the conception and development of practices reflect the very nature of the food systems, in which feedbacks and factors of an economic, social, and environmental nature operate. In fact, the competition for resources (primarily the soil) and between man and the environment (including fauna) are among the main causes that gave rise to the projects, together with the territorial imbalances that these frictions trigger. Social factors—the lack of social capital and mutual trust between territorial actors—and the economic one—lack of support for sustainable supply chains and networks of producers and consumers—make up the framework of the critical issues from which the practices arose. Following the planning and in conjunction with the application of the practices, two types of factors emerged: on the one hand, some elements of success were identified by the representatives interviewed; on the other, some obstacles and critical factors (both endogenous and exogenous) have imposed a review and adaptation of the project. Among the former, good project management (constant coordination, good identification of the problem, strong motivation and involvement of the partners), the activation of institutional and civil society actors, the empowerment of the partners, and the involvement of the communities were noted, as well as the recovery of ecological functions and the birth of new business opportunities. Among the obstacles the main categories that affected the project were the balance and political relationships, the lack of participation of some stakeholders or the conflict between some partners, the dependence on established paths and practices, hardly questioned (preconceptions, resistance bureaucratic, understanding of potential), and the endogenous and exogenous difficulties of the project (insufficient resources, changing environmental conditions, technical/administrative difficulties). It is important to underline how the obstacles encountered during the project derive from the pressures caused by the fact that the change proposed by the practices activates feedback from both the environmental and socioeconomic systems. Added to this are the pressures, factors, and conditions deriving from the socioecological context of the territory in which the project intervenes. To respond to these obstacles that occurred during the project, some solutions were identified and put in place by partners active in resilience practices. In particular, the strengthening of the training of the partners, the efforts on project communication, the design transparency, and the facilitation of relationships were among the strengths to overcome the obstacles encountered. The ability to be flexible and to change the way the project is implemented is among the factors that have enabled the success of the initiatives. Finally, the construction and facilitation of networks, the stimulus for participation, the organization of events, and the concreteness of the actions have allowed project leaders to overcome the obstacles linked to mistrust and the difficulty in understanding the usefulness of the project. Some unexpected impacts, or the extent of which had not been foreseen before the start of the project, were finally reported by the interviewees. These concern the increase of knowledge and awareness on the topics covered by the intervention, both by institutional actors and by the civil society involved. Furthermore, some projects had the ability to create the conditions for an autonomous continuation of the activities even after the project financing phase, also thanks to the activation of interested neighboring actors. The ability to disseminate practices and objectives even at higher or wider territorial levels than those envisaged by the practice is among the most important institutional and governance impacts. Finally, the activation of further projects and financing following the conclusion of the practices has allowed some practices to involve a greater number of stakeholders, to guarantee the economic and financial sustainability of the activities, and to give continuity to the project.

7.5 Conclusions

The chapter analyzed, 5 years after their launch, a set of local practices that joined the ORP, a project started in 2015, funded by the Cariplo Foundation, and conducted by the Politecnico di Milano. The ORP collected practices that aimed at strengthening territorial resilience on the different dimensions of sustainability (social, cultural, environmental, economic). In particular, practices that directly or indirectly act on local food systems have been extracted from the observatory. The theoretical and methodological background used is that related to SESs, in which the interactions between man and the environment are considered a coupled system in which socioeconomic and biophysical driving forces interact to influence food system (and subsystem) activities and outcomes, both of which subsequently influence the driving forces. The results of the survey highlighted the interactions between the drivers that motivated the emergence of practices, the success factors, the exogenous and endogenous obstacles that occurred during the projects, the solutions adopted and the unexpected impacts. These elements allowed to build a model that represents the flow of these interactions, which was elaborated in Fig. 7.1. The research contains the ex post evaluation of some resilience practices and represents the conclusion of a research path that began with an ex ante analysis of 50 local practices related to food–city relationships. Overall, the chapter helped highlight some connections and feedback that were activated during the implementation of the project, creating an analysis and evaluation model that can be replicated for other resilience practices.

References

Adger, W. N. (2006). Vulnerability. *Global Environmental Change, 16*(3), 268–281.

Dezio, C., & Marino, D. (2018). Towards an impact evaluation framework to measure urban resilience in food practices. *Sustainability, 10*(6), 1–13.

European Commission. (2017). *The future of food and farming; communication from the commission to the European parliament, the council, the European economic and social committee and the committee of the regions*. Brussels: European Commission.

FAO. (2017) The State of Food and Agriculture 2017. Leveraging food systems for inclusive rural transformation. Available from http://www.fao.org/3/a-I7658e.pdf (accessed on January 2018).

Foran, T., Butler, J. R., Williams, L. J., Wanjura, W. J., Hall, A., Carter, L., & Carberry, P. S. (2014). Taking complexity in food systems seriously: An interdisciplinary analysis. *World Development, 61*, 85–101.

Holling, C. S. (1973). Resilience and stability of ecological systems. *Annual Review of Ecology and Systematics, 4*, 1–23.

Kallis, G., & Norgaard, R. B. (2010). Coevolutionary ecological economics. *Ecological Economics: the Journal of the International Society for Ecological Economics, 69*, 690–699.

Marino, D., Mastronardi, L., Giannelli, A., Giaccio, V., & Mazzocchi, G. (2018). Territorialisation dynamics for Italian farms adhering to alternative food networks. In M. Dymitrow, & K. Halfacree (Eds.), *Bulletin of Geography. Socio-economic Series, No. 40*. Available from http://doi.org/10.2478/bog-2018-0018.

Marino, D., & Cavallo, A. (a cura di), (2014) Agricoltura, cibo e città. Verso Sistemi Socioecologici Resilienti; CURSA (pas) SAGGI—Ricerche e Progetti: Roma, Italy, 2014; ISSN 2284-4376.

MEA. (2005). *Millennium Ecosystem Assessment: Ecosystems and Human Well-Being: Synthesis*. Washington, DC, USA: Island Press.

Norgaard, R. (1984). *Coevolutionary Agricultural Development. Economic Development and Cultural Change* (32, pp. 525–546). Chicago, IL, USA: The University of Chicago Press.

Petrosillo, I., Aretano, R., & Zurlini, G. (2015). Socioecological systems. In S. A. Elias (Ed.), *Reference Module in Earth Systems and Environmental Sciences*. Elsevier. Available from http://doi.org/10.1016/B978-0-12-409548-9.09518-X.

Prosperi, P., Allen, T., & Cogill, B. (2019). Building consensus on sustainable food system assessment. In A. Blay-Palmer, D. Conaré, K. Meter, A. Di Battista, & C. Johnston (Eds.), *Sustainable Food System Assessment: Lessons from Global Practice*. London: Routledge. Available from https://doi.org/10.4324/9780429439896.

Steffen, W., Richardson, K., Rockström, J., Cornell, S. E., Fetzer, I., Bennett, E. M., & Biggs, R. (2015). Planetary boundaries: Guiding human development on a changing planet. *Science (New York, N.Y.)*, 347. Available from https://doi.org/10.1126/science.1259855.

Turner, B. L., Kasperson, R. E., Matson, P. A., McCarthy, J. J., Corell, R. W., Christensen, L., Eckley, N., Kasperson, J. X., Luers, A., Martello, M. L., Polsky, C., Pulsipher, A., & Schiller, A. (2003). A framework for vulnerability analysis in sustainability science. *Proceedings of the National Academy of Sciences, 100*(14), 8074–8079.

Walker, B., Holling, C. S., Carpenter, S. R., & Kinzig., A. (2004). Resilience, adaptability and transformability in social–ecological systems. *Ecology and Society, 9*(2), 5. Available from http://www.ecologyandsociety.org/vol9/iss2/art5/.

CHAPTER 8

Resilience Practices Observatory project: synergies for stabilization and upscaling

Giulia Pesaro and Angela Colucci
REsilienceLAB/Co.O.Pe.Ra.Te. Ldt, Italy

The success of the Resilience Practices Observatory (RPO) project demonstrated the *resilience practices'* need for a shared space of networking, dialog, and learning. In 5 years of activity (2015–19), the RPO engaged more than 100 different community-led practices acting on the local resilience capacity improvement.

RPO provided a shared space to resilience practices supporting their visibility, promoting experiences exchange and learning, and a common framework (and interdisciplinary competencies) to think to "individual" experience in terms of contributions to a more complex vision and scales. The RPO played a unique role with peculiar characteristics, becoming a space to understand the role and contributions of resilience practices not locally and for interventions promoted but also in terms of complex processes involving and providing multiple benefits to environmental, social, economic, and organizational components of territorial systems.

The chapter aims to retrace the main core issues emerging from the practices' geographies, address contributions in the debate on grassroots practices, make explicit success factors of the RPO experience, and outline transferability perspectives (or unicity) from lessons earned.

From the diversified activities related to the Resilience Practices Forum (RPF) meetings and discussions, three main domains emerged as barriers and, at the same time, as fundamental aspects to support the feasibility, success, and stabilization of practices. The emerging barriers also refer to a more general growth and improvement of the cultural conditions needed to stabilize and diffuse strategies toward the reinforcement of resilience proprieties of complex systems and the empowerment of communities, integrating regional strategies/policies.

[ECO]systems of Resilience Practices
DOI: https://doi.org/10.1016/B978-0-12-819198-9.00012-0

The weaknesses and barriers to be overcome emerged from the capacity-building activities organized involving several actors: the coordinator of resilience practices, institutional stakeholders, private and economic actors, academics, and experts engaged. The main weaknesses could be clustered in three principal assets:

- process weakness and barriers that include both exogenous factors (such as difficult relationships with already existing practices and processes, difficulties in the building of local and over local networks) and endogenous or strongly locally based (weaknesses in the processes to involve and make all elements of the target communities participate to the project design and implementation phases; cultural barriers and mistrust in innovative tools, mainly the sharing and cooperative ones, needed to operationally produce the active involvement which in many cases is the crucial element for the practice success);
- administrative and bureaucratic weaknesses producing timing problems and obstacles in decision-making and activities needed to implement the practices. Diffuse scarce confidence in the public institution and distrust in the feasibility and effectiveness of public—private cooperation emerged (and it is an urgent aspect to be addressed);
- endogenous barriers for stabilization refer primarily to the economical sustainability of both practices and local context (lack of financial means or stability and continuity of the funding).

8.1 Enabling capacities for stabilization

The urgency of innovation and renovation of all stakeholders' capacity-building and learning processes involved in the resilience practices processes emerged along with all the RPO project activities. The innovation is needed to promote virtuous processes of incremental learning of all stakeholders involved (academic, experts, institutional, operative, citizen) (Chaskin et al., 2017). Consolidated methods were integrated with innovative ones, engaging different stakeholders' categories, developing activities rooted in the local context, and with local communities in an active and proactive role.

Improving the capacities of the resilience practices emerged both as a crucial aim solicited in all the RPO activities and as an implicit asset of practices' actions itself. In particular, the urgency innovating capacity-building tool of all stakeholders involved in the resilience practices processes emerged from the RPF process. The innovation is needed to promote virtuous processes of incremental learning of all stakeholders involved (academic, experts, institutional, operative, citizen). The RPO activated capacity-building initiatives and learning based on consolidated methods integrated with innovative ones to engage different stakeholders, rooting capacities on real and local-based challenges, promoting an active role and the actors' commitment. Increasing the capacities of the actors who promote and manage the practices can guarantee stabilization of the processes activated and implement the actions promoted. Improving the capacities of all the actors involved in the process is necessary to ensure that the "ecosystem" of practices can give systemic contributions and benefits to local territories (or at larger scales). Based on the RPO experience evidence, it is possible to affirm that facilitating multiple benefits is crucial to improve the capacities of the resilience practices lead-actors and all the stakeholders involved in the transformation process, from institutional actors to academic and researchers, toward an effective innovation in processes and in multistakeholder governance.

Analyzing the resilience practices experiences, capacity-building and learning emerge as core and diffuse action domains of practices that activate different capacity-building tools. In the first phases of the process, the resilience practices (involved and analyzed) activated capacity-building tools to improve their capacities to manage the implementation process (as organizational, other specific knowledge and tools like Information and Communications Technology). Tools of capacitation are also the core domain of action of several resilience practices that launched specific initiatives of capacity-building to enhance the capacities of targeted stakeholders engaged/involved in the project implementation (and goals achievement). Several resilience practices focusing on sustainable local food systems analyzed in Chapter 7 exemplify this double declination of capacity-building tools. As the authors underline, the analyzed practices are aware of needing an internal capacity-building to manage the project's implementation and activate capacity-building actions as a core tool to promote innovations toward the desired sustainable local food system. In the first phase the practices launched activities centered on internal capacity-building (non-governmental organization (NGO) members, actors of the partnerships) mainly oriented to improve organizational, ICT, or other missing capacities for the management of the process. In the subsequent phases the practices activated both capacity-building tools to improve the citizen's awareness toward more sustainable food behavior (such as orienting the individual purchase toward local, sustainable food market, toward more sustainable dietary) and capacity-building tools to improve the capacities of local farmers (private actors) in reorganizing the production models, in accessing to local (private/public) markets.

Capacity-building is often used as an undefined and umbrella concept containing a broad range of tools that are very different in relation to the scope and the target of the capacitation tool (Chaskin et al., 2017). Concerning the practices experiences analyzed, capacity-building tools could be clustered concerning the scope of the tool itself:

- capacity-building tools for awareness and knowledge improvement for the diffusion of urban climate issues in large cultural social–cultural landscape (Galderisi & Colucci, 2018);
- tools for capacities enhancement to enable/improve the implementation, realization of strategies, and actions/initiatives (Frantzeskaki & Rok, 2018; Militaru, 2018).

The main scope of tools for awareness improvement is to improve the awareness on resilience topic to reach a diffuse culture with the final scope of orient individual behaviors modification of citizens and local communities and reach a mainstreaming of resilience principles in public decision-making. The tools for knowledge improvement are mostly connected to disseminating the climate change issue and other challenges concerning the resilience enhancement of complex systems (e.g., scientific knowledge transfer and dissemination, including the scientific knowledge in the "cultural" and political sphere).

Focusing on climate change issues, the tools for awareness and knowledge improvement aim to shift the sociocultural model (transition) to a more sustainable ones and to orient the integration of the climate change issues every day (citizens behavior) and political decision-making. For example, a capacity-building tool can improve citizens' awareness in adopting more sustainable behavior reducing the impact of our everyday action (such as slow mobility, energy transition, water management, sustainable/healthy food consumption). Awareness and knowledge capacity-building tools could be targeted. Both capacity-building tools awareness and knowledge could be targeted to community/citizens, technical staff, and practitioners to improve/orient the cultural awareness and impact on political–cultural level (Loorbach, 2010).

Another category of capacity-building tools is addressed to improve the capacities (of stakeholders) to enable the implementation of resilience practices interventions. The capacity-building tools for improving action capacities can be addressed on the "technical knowledge" in improving the capacities in managing the implementation/management of urban climate change solutions (targets: civil society organizations, practitioners, technical staff of public and private bodies). The capacity-building tools' scope provides capacities/skills needed (or lacking) to guarantee the climate change policies implementation and stabilization (and long-term success).

If resilience practices engaged demonstrated to have in capacities improvement as core domains of their action, it is also crucial to underline as the generic *capacities-building* term is often referred to an extensive and undefined range of tools. The capacity-building tools are potentially crucial in climate adaptation and sustainability policies/actions implementation not effective if not correctly planned and designed. A single capacity-building tool must be designed concerning the clear identification of the scope (awareness, knowledge, capacities) and the clear identification of the target.

Several practices are rooted in the partnership between academic/research and civil society actors. However, during the RPF (the 2018 RPF, mainly dedicated to hybridization), the actors noticed a gap and communication barriers with the research and academic world. These difficulties were addressed concerning three principal aspects:

- the approach of academic/research actors often felt like a "lesson" and not as a mutual process of coproduction and synergy,
- the practices "lives" the academic contribution when characterized by excessive specialization and distance from the operational dimension like a sort of "loss" of time compared to the energy to be dedicated to the action, and
- excessive specialization of scientific knowledge often implies difficulty in translating excessively sectorial aspects into operations.

These emerging affirmations lead to two considerations. First, a widespread renewal of "academia" is undoubtedly necessary in approaching with a collaborative and equal disposition toward knowledge exchange paths (different but present both in the academic–scientific world and in the practices of experiential and operational knowledge). Second, it again proposes a well-known leitmotif that knowledge and complexity are a waste of time while understanding complexity and investing time in constructing the process. Long-term strategies must become an essential step of the value of practices to ensure stabilization and maximize multiple benefits.

From the sustainability paradigm the world of academia and research has made considerable efforts to increase the skills in sharing knowledge and raising awareness and scientific communication, bringing notable advances. However, academia and research still have to improve the capacities in knowledge coproduction to include and stimulate knowledge to overcome mistrust in the value of time invested in understanding and designing complex phenomena.

Excessive specialization is a recurring barrier aspect of territorial policies and practices and constitutes a relevant limiting aspect of the effective implementation and activation of integrated and multiobjective strategies and processes.

8.2 Knowledge hybridization

The hybridization among sectors and disciplines is a well-known issue in research innovation and territorial transformation management and is needed to approach and overcome witched challenges and complex phenomena. The resilience approach boosts the synergy and interdisciplinary approach based on the socio-ecological system (Berkes et al., 2000; Folke & Gunderson, 2010; Folke, 2006).

Academic actors are committed to improving their capacities to contribute to the transition cultural models toward sustainable and resilient scenarios. Hybridization among epistemologies, including local/material knowledge and different epistemology domains science, art, and human, appears as crucial strategic tools in boosting transformative cultural transitions. The knowledge issue is a crucial aspect of the transformation of complex systems through the resilience lenses. Several resilience practices in international and RPO panorama have as fulcra the individual and communities' skills (recognition and sharing) and take the human and social resources as values and pillars for the "action." Literature (and applied research) developed this topic in terms of coproduction. In particular, coproduction refers to actors' active involvement and engagement in producing knowledge inside decision-making processes to accomplish an effective engagement of stakeholders (Voorberg et al., 2015). Research highlighted that bringing together actors from multiple disciplines and diverse competencies and backgrounds in an open process for sharing and learning is a precondition for cocreating new knowledge for problem-solving (Rotmans et al., 2001). At the same time, research demonstrated that coproduction and shared knowledge processes are resources in developing innovation and advancements in academic and scientific knowledge (Frantzeskaki & Rok, 2018) strengthening the multidisciplinary elements and the "abilities" to recognize and activate connections among a variety of sectors and their characteristics and phenomena.

Based on the experience of the RPF process, three main issues emerged:

- The demand for and the opportunity to develop models promote dialogs and integrating multiple types of knowledge. The need to integrate different types of "disciplinary" knowledge is evident in developing transdisciplinary research. The RPO experience demonstrated that the dialog and exchange among different types of knowledge (scientific/expert knowledge, tacit knowledge, institutional knowledge, "practices" knowledge, etc.) could ensure innovation and creativity.
- Models based on multiple types of knowledge have to be also developed in the "early" steps of resilience practices processes, starting from the activation of the process, the problem setting, and the characterization of the local system.
- Resilience approaches imply innovation in understanding the context through expert/scientific knowledge and the ability to identify all existing resources and opportunities. The identification of opportunities and resources results from "recognized links" between separated phenomena and the integration of sectorial competencies and multiple knowledge sources.
- The cultural process about resilience: The cultural process is fundamental to reaching a real implementation of the adaptation process and coping and dealing with dynamic equilibria and uncertainty. The cultural process is necessary for implementing innovative "tools" in terms of methods, design and planning solutions, and policies. It is essential to act on the cultural process to promote and activate differentiate learning initiatives on different resilience approaches (from the more consolidated talks to liv-lab activities and experiments) (Bulkeley et al., 2016).

RPO's role was capacitation of all stakeholders. The resilience practices need to be based on the constant and incremental learning process of all involved actors and processes of knowledge coproduction, involving multiple sectors and sources, preparing scientific, expert, and local communities to understand the dynamic phenomena and to cope and adapt with uncertainty, activating complex strategies also based on psychological and social components.

8.3 Ecosystem of practices: networking and governance process of Resilience Practices Observatory practices

The term governance refers to all differentiated mechanisms (self-regulation or informal rules of civil society, coregulation of public and private actors, and regulation through institutional government) through which individuals and institutions, public and private actors collaborate in managing common interest and social affairs (Duit et al., 2010; Salamon & Elliott, 2002).

From the dialogs activated by RPO the lack of intentional "design" of the governance process emerged, as well as it was clear the need for capacity-building and awareness on this aspect to reach a stabilization (and improve the benefits to local/global systems) of resilience practices. These also imply the previously introduced renovation of toolboxes, according to the innovative environment which characterizes the resilience practices: innovation in knowledge (co)production, in the decision-making process, in design and solutions identification (from "physical" and technical solutions to more complex and integrated strategies), implementation and management of initiatives (engagement and empowerment of all actors).

From the RPO activities and database, some general models of processes governance emerged that also are recognizable in the international panorama of resilience practices and programs:

- An institutional leadership characterizes the climate change adaptation and prevention risk programs and plans: adaptation plans are usually promoted by metropolitan or city government. The whole of the processes promoted in the practices are voluntary, strategic, and mainly characterized by a large range of participatory and inclusive paths, to involve and engage various stakeholders and include existing initiatives.
- Bottom-up processes characterize transition initiatives that are activated, promoted, and implemented through the mobilization and actions of citizens. A key factor for the success of the projects, in terms of consolidation and continuity of initiatives, is related to setting up a solid and rich partnership around the initiative (with the presence of associations, NGOs, private sector, and public/institutions).

The RPF and RPO activities revealed a panorama showing the diffusion of new governance models not included in the consolidated framework of regional and planning governance instruments (Bulkeley et al., 2016). Inspiring concepts and proposals were developed in relation to metropolitan and urban regeneration processes, integrating environmental and ecosystem issues with social and economic innovation in terms of crosscutting and integrated solutions and new models of governance. These dynamics recall the concept of adaptive governance (Olsson et al., 2006) and reflexive governance (Chapter 16) and can be seen as examples of innovative approaches to governance, where a mix of different tools can be found for the stakeholders' engagement and empowerment.

The relevance of investigating (with renovated lens) the typical dichotomy of categories labeled as top-down and bottom-up emerges from the RPF practices panorama. Some practices identified as top-down related to climate change adaptation and risk prevention are characterized by a complex governance process able to engage in the decision-making process an extensive range of stakeholders (remains the fact that governance processes are not homogeneous observing the panorama of climate change adaptation programs). At the same time, investigating the quality of the governance process of initiatives promoted by local committees and associations (citizens, associations) showed clearly as the level of efficacy and inclusivity of bottom-up processes is not homogeneously presenting both open processes and several cases of uninclusive process of decision-making (close process).

During the RPF the engagement of citizens, communities, and other territorial actors emerged as one of the most critical barriers for the implementation of the bottom-up practices panorama. Concerning the citizen involvement and communities-target engagement, it is possible to underline a general lack of awareness in the definition (or design) of the actors' "map" for the governance process: a clear stakeholder's identification is a crucial aspect to reach the quality of governance, and often coordinators of practices are used to maintain consolidated alliances. In these terms a lack of an explicit involvement of stakeholders emerged. Private or public bodies supporting the implementation through economic grant/fund and public and private actors working on the innovation like universities, research centers, and consulting companies could be partners (Creech et al., 2008). The database analysis also permits to affirm that the capacities of resilience practices in setting mixed partnerships enable more generative processes. At the same time, the RPO project showed how a siloed approach could also characterize resilience practices led by no-profit actors: difficulties in activating crosscutting connections and in integrating different initiatives, policies, and projects related to different sectors or spheres affect in different ways a large number of resilience practices. The lack of virtuous contaminations and alliances between sectors and actors is a common critical issue for processes/practices implementation managed by different typologies of governance models: the processes promoted by public/institutional, private, and civil society and/or addressed to different sectors often are not converging (Ravi, 2019).

The networking and the synergies could be recognized as crucial factors in stabilizing community-led processes otherwise fragmented, fragile (Colucci, 2018). The networking permits to share both competencies and built new alliances. Networks of practices can act (Ostrom, 2009) as the third actor in decision-making processes. At the same time the definition of networking is a complex process: among practices that adhere to RPO, some are part of national or international networks that are often characterized by themes (food, agroecology, social innovation, etc.). Thematic networking is widespread and growing because it provides an advantage for sharing operational tools (knowledge and skills) for operations. At the same time, thematic networks do not support the comparison and hybridization between components (environmental, social, economic, etc.) and fields of action that often characterize third sector actors. An unsolved question is the role (who) of community-led networking is discussed in the chapter on adaptive governance (Chapter 16 focuses on networking and governance topic).

The RPO could be defined as a "networking" space, in a unique position to be a third and hybrid partnership able to boost and promote the synergies activation among practices and institutional, academic, and private stakeholders. Networking is not based on one topic but inviting and open to the topic and resilience approaches to share in a mutual learning process from experiences developed. RPO played as a unique third space connecting institution, market, research, and "practices" (or common referring to Ostrom, 2009).

8.4 Economical stabilization sustainability and independency

There are many reasons why economic thinking approaches, factors, criteria, and toolboxes should be better considered potentials in resilience practices and in the related project design processes. The reading of the practices collected through the RPO and the RPF demonstrated that, when working on projects for developing new practices aimed at enhancing the resilience of urban areas, territories, and communities, the integration of an "economic-based" reasoning could highly contribute to enhance the project itself for some reasons.

- It is necessary to be more aware of how to deal with the economic and financial components integrated in the projects. Such components exist in quite all cases but many times remain in a background position, while the economic dimensions could be responsible for the failure of a project over time, especially over a long-term horizon.
- It is necessary to increase the quality and the innovation contents of the economic- and financial-based tools available. These both in the design processes and in the upstream stage, where public and private institutional and administrative bodies take decisions about themes and sectors which will be funded and about the set of rules and criteria to be applied in the selection of projects.
- There are increasing opportunities in better integrating projects on urban territorial resilience with methodological approaches and action goals looking at their economic dimensions. This is particularly when referring to the more innovative tools and practices available in the so-called new economies, based on collaboration, sharing, and sustainability.

Going deeper inside these points, the first reference is to the project design processes of the resilience practices, where the capability to better integrate the economic components and stabilize them over time is an emerging issue. This of course regards the economic and financial feasibility of the project implementation and maintenance and the direct economic revenues estimated by/for the stakeholders and participants, if expected.

At the same time, economic concerns regard the sustainability itself of the projects in the medium and long terms and the related positive externalities, recognizable in the values/advantages produced for a broader community than the target one. This means that the practices implementation produces advantages not only on the community or on the territorial areas directly involved in the project but also at a broader level, due to a positive territorial cumulative effect.

The reading of the practices demonstrates how difficult it is to design economic elements in resilience practices. This seems due to two different effects. On the one side, it seems that working on resilience does not consider as a main potential goal the economy and the economic values. In some cases the economic feasibility dimension itself is really poor, even in funded projects. On another side, in project design teams the economic culture and skills are many times lacking, which explains in many cases the reasons of the poor economic contents. The better integration of the economic thinking toolbox in resilience practices needs the involvement of, let us say, "new interdisciplinary experts," able to make the economic matters one of the possible key issues in resilience practices.

The reading key here is different from the more diffused approaches dealing with *resilient economy* or *resilience and economy* and with the ways by which it is possible to measure the resiliency of an economic system (Briguglio et al., 2009). The need for an enhancement in the participation of economic thinking experts in projects design teams is mainly linked to the understanding and the better "use" of the economic components as allies toward more resilient urban and territorial communities (Pesaro, 2015).

To facilitate this process, economic components should be better seen in their enlarged meaning, that is a system of resources and tools able to produce values that cannot every time be measured in a quantitative way (money) and are therefore more difficult to be perceived. It does not mean that they do not exist but need to be better explained and made visible to communities, for example, by implementing specific cultural paths.

From these the second thinking element comes. The integration of the economic thinking and toolbox, together with the idea of a capital of territorial resources to be regarded as a set of opportunities, of course in a sustainable development perspective, could produce innovation in the practices design and implementation tools. The innovation itself could be found in new, more sustainable and effective ways to use the territorial capital of resources, both material and immaterial ones. Beyond direct economic feasibility elements and looking toward the stabilization of resilience practices to make them become "ordinary," innovation in action models could rely on a different way to work on the whole of the local resources, looking at the potential value flows coming from their use. Of course, this is possible referring to the more innovative frontiers of the economic thinking, based on the preservation of the territorial capital/heritage, on the maximization of the value flows, and on a different perspective in the assessment of values. This is especially when innovative goods and services are at stake, and it is difficult to evaluate the related values.

Such a process of innovation is very important if we consider the funding processes. In the enhancement of the community and territorial resilience being a continuously evolving approach to territorial development (as the main international activities show), the practice diffusion still needs a lot of financial support to be implemented. The role of funding bodies is therefore more and more crucial for the enhancement of the system.

The so-called funding bodies are all the public and private institutions who fund, in different ways and inside different programs and calls, the diffusion and implementation of the resilience practices. This means that their contribution and approaches are crucial not only because of the funding amounts but mainly in their ways of "shaping" what can be regarded as the "resilience in action." This is because such bodies select the projects by identifying the implementation sectors, the eligible subjects, the goals set, and the evaluation criteria.

This is true in many fields linked to the protection and production of "commons," where funding still remains the main economic source to activate communities and territorial subjects to produce public goods and services. We can therefore consider resilience as a public good, like safety or education (Pesaro, 2007) and the funding resources as an "economic element," even if indirect, which becomes a key to the understanding and diffusion of the practices. The role of the funding bodies is therefore strengthened, together with their responsibilities in the enhancement of resilience as a quality of a community or of a territorial area.

Another key element is that the same bodies, by their selection processes, might contribute to enhance innovation in action models and practices and to stabilize the system over time. Stabilization, in particular, seems crucial to let the territorial systems "produce resilience" in a long-time perspective. Among others, this means to continue a certain funding program for some years, like in the case of Fondazione Cariplo (the same call has been repeated for 4 years) to produces synergies among practices, let the territorial stakeholder take advantage of the offer and begin to see visible value effects "on the ground." These are "economic words," by which one can express a potential in enhancing effectiveness and efficiency of the funding program and of the related results. Resilience, like sustainability, is a long-term goal by nature and is enhanced by the continuous, transversal, and diffused action, with a cumulative effect in results.

At the system level, this suggests aiming for a stronger coordinated action. Funding bodies and institutions should then better look to the development of "metasystemic" and strategic projects, both public and private institutional, within which even the small size and short-term projects can find greater meaning as components of a "designed system." On the one hand, the stabilization of funding opportunities could create the conditions for a continuation of actions and interventions that can better support the completion of the projects and the related implementation. On the other hand, this would allow to place/offer to territories more comprehensive and ambitious resilience and systemic sustainability goals.

Regarding the latter point, in the light of the socioeconomic context that characterizes our territories, the issue of economic resilience itself of territories and local communities has to be carefully considered. This could become a system-level goal of action, a general macrolevel reference for the medium and long terms in a development perspective. A systemic goal, by identifying some possible direct results, can help give greater coherence and to address the set of sectoral interventions in the territory, both small and great they might be in terms of investments, involved subjects, operational objectives, and expected benefits. What is important is not the size but the internal consistency, necessary, and central to converge on and help achieve common goals through a sequence of short-term activities, provided that they are more consistent and repeatable over time.

The increasingly important contribution of the tools and of the methodologies characterizing the so-called new economies comes into play in a more explicit way. The new economies, based on principles like *smart, sharing, collaborative, circular,* and *green* can represent the best possible "partners approaches" to enhance the economic side of the resilience practices. This, being aware that the expected changes and results are new or renewed sets of economic goals for a territorial system and deeply involve the system itself of the production and consumption models, which are expected to become consistent with a sustainable future.

The attention to the economic components, in all of these meanings, appears then beneficial and useful, an innovative interpretation key for the evaluation of the resources and forces of a territorial system. A key able to address the whole of the local activities toward new structural equilibriums pays more attention to quality and long-term sustainability goals.

8.5 *Resilience Practices Observatory*: contribution to resilience enhancement

The previous paragraphs explored four/three aspects that emerged to stabilize resilience practices and practice contribution. This paragraph aims to discuss the multiple contributions from the practices and some concluding remarks on the experience of the RPO focusing on the general contributions of practices in strengthening the resilience capacities of urban and territorial systems.

First, the shared and crosscutting core asset of action of resilient practices is the commons and local/global sustainability. The urgency and levers for resilience practices activation intercept an extensive range of topics, but all practices acting on common are defined by Ostrom (1990).

RPO experience demonstrated both the crosscutting attention of practices (across challenges and topics) on capacity-building tools that are often recognized as a core focus and the integration of social dimension in the resilience practices processes even if rooted in environmental issues.

For the most part, RPO practices have been shown to activate incremental processes of learning and capacitation. This aspect, demonstrated both by the relevance of actions of capacity-building and learning as the heart of the action of practices and by the participation and activation of practices in the capacitation activities of RPO, shows a widespread maturity of the actors' promoters of practices that are aware of the need for continuous improvement of skills and with an attitude aware of internal limits and skills (missing) necessary to implement the action/project.

At the same time, the capacity-building tools devoted to local communities' awareness are often, if not the principal purpose, an action domain included in almost all the practices analyzed. Improving the awareness of communities and citizens is a relevant result to be recognized in resilience practices. Practices, acting, and activating dialogs at local scales contribute to the mainstreaming process and the cultural advancement of resilience, sustainability, adaptation, and climate issues and principles in an extensive range of sectors and actors starting from the citizens and communities.

Another crucial contribution is in boosting the integrated approach. As presented in Chapter 5, practices activate complex processes articulated in multiple actions and interventions synergic to environmental, social, and economical sustainability, demonstrating the indissoluble connection among environmental emergencies, social and economic challenges in local action. For this strong integration among typical environmental (climate, risk, landscape) challenges and social inclusion issues, the practices are often recognized as an inspiration for innovation in the transformative model of our territorial and urban context from extensive academic research.

Since last decades academic and disciplinary debates and research have increasingly paid attention to transition initiatives as spontaneous intervention and grassroots innovations and their growth, both concerning geographical diffusion and issues of interest and action and spread. Differentiated initiatives activated and managed by local groups acting on the physical dimension and the immaterial dimension of commons became a relevant worldwide phenomenon. Also, in planning, urban design, and urban policies debate, their role became more central (Banerjee, 2001). Several research and projects have recently proposed renovated approaches to urban transformation in planning and urban policies, integrating and valorizing energies and innovations emerging from spontaneous initiatives and based on more robust participative and inclusive codesign processes. An example of this attention is demonstrated, for example, by the US pavilion presented in 2012 at Biennale di Venezia event where "Spontaneous Interventions: Design Actions for the Common Good" celebrated and presented initiatives of citizens able to transform critical urgencies of urban context into new opportunities to become back to public life.

In literature, debate transition initiatives are investigated through a differentiated lens concerning innovative and creative proposals and solutions (grassroots innovations, social innovation), concerning community-led governance processes (coproduction and codesign, participative urban interventions), concerning social aspects (commons and public services, urban and public policies).

Resilience practices, embracing and including an extensive range of local-based processes led by citizens groups (NGOs) and private sectors and experts, demonstrated their impact in addressing local issues from social innovation to sustainable local behavior and social inclusion and contributing to sustainability goals. The practices activate often processes balancing local context urgencies and community priorities of action to deal with local dynamics and global phenomena. The practices demonstrated their contribution in connecting different challenges and local - territorial and global scales.

Attribution

Although the chapter is the result of a common reflection, Sections 8.1–8.3 and 8.5 have been written by Angela Colucci, and Section 8.4 has been written by Giulia Pesaro. The introduction is the result of common contribution.

References

Banerjee, T. (2001). The future of public space: Beyond invented streets and reinvented places. *Journal of the American Planning Association, 67*(1), 9–24. Available from https://doi.org/10.1080/01944360108976352.

Berkes, F., Folke, C., & Colding, J. (2000). *Linking Social and Ecological Systems: Management Practices and Social Mechanisms for Building Resilience*. Cambridge University Press.

Briguglio, L., Cordina, G., Farrugia, N., & Vella, S. (2009). Economic vulnerability and resilience: Concepts and measurements. *Oxford Development Studies, 37*(3), 229–247. Available from https://doi.org/10.1080/13600810903089893.

Bulkeley, H., Coenen, L., Frantzeskaki, N., Hartmann, C., Kronsell, A., Mai, L., Marvin, S., McCormick, K., van Steenbergen, F., & Voytenko Palgan, Y. (2016). Urban living labs: Governing urban sustainability transitions. *Current Opinion in Environmental Sustainability, 22*, 13–17. Available from https://doi.org/10.1016/j.cosust.2017.02.003.

Chaskin, R. J., Brown, P., Venkatesh, S., & Vidal, A. (2017). *Building Community Capacity* (1st ed.). Routledge. Available from https://doi.org/10.4324/9781315081892.

Colucci, A. (2018). *The transition approach in urban innovations: Local responses to climate change. Smart, Resilient and Transition Cities: Emerging Approaches and Tools for a Climate-Sensitive Urban Development*. Elsevier.

Creech, H., Vetter, T., Matus, K. J. M., & Seymour, I. R. (2008). *The governance of non-legal entities: An exploration into the challenges facing collaborative, multistakeholder enterprises that are hosted by institutions*. International Institute for Sustainable Development. Available from http://www.iisd.org/publications/pub.aspx?id = 1044.

Duit, A., Galaz, V., Eckerberg, K., & Ebbesson, J. (2010). Governance, complexity, and resilience. *Global Environmental Change, 20*(3), 363–368. Available from https://doi.org/10.1016/j.gloenvcha.2010.04.006.

Folke, C. (2006). Resilience: The emergence of a perspective for social–ecological systems analyses. *Global Environmental Change, 16*(3), 253–267. Available from https://doi.org/10.1016/j.gloenvcha.2006.04.002.

Folke, C., & Gunderson, L. (2010). Resilience and global sustainability. *Ecology and Society, 15*(4), art43. Available from https://doi.org/10.5751/ES-03954-150443.

Frantzeskaki, N., & Rok, A. (2018). Co-producing urban sustainability transitions knowledge with community, policy and science. *Environmental Innovation and Societal Transitions, 29*, 47–51. Available from https://doi.org/10.1016/j.eist.2018.08.001.

Galderisi, A., & Colucci, A. (2018). *Future perspectives. Smart, Resilient and Transition Cities* (pp. 285–292). Elsevier. Available from https://doi.org/10.1016/B978-0-12-811477-3.00034-1.

Loorbach, D. (2010). Transition management for sustainable development: A prescriptive, complexity-based. *Governance Framework. Governance, 23*(1), 161–183. Available from https://doi.org/10.1111/j.1468-0491.2009.01471.x.

Militaru, A. M. (2018). Building capacity for sustainable development through community-based education. In W. Leal Filho, M. Mifsud, & P. Pace (Eds.), *Handbook of Lifelong Learning for Sustainable Development* (pp. 229–241). Springer International Publishing. Available from https://doi.org/10.1007/978-3-319-63534-7_16.

Olsson, P., Gunderson, L. H., Carpenter, S. R., Ryan, P., Lebel, L., Folke, C., & Holling, C. S. (2006). Shooting the rapids: Navigating transitions to adaptive governance of social-ecological systems. *Ecology and Society*, *11*(1), 18.

Ostrom, E. (1990). *Governing the Commons: The Evolution of Institutions for Collective Action* (1st ed.). Cambridge University Press. Available from https://doi.org/10.1017/CBO9780511807763.

Ostrom, E. (2009). *Understanding Institutional Diversity*. Princeton University Press.

Pesaro, G. (2007). Prevention and mitigation of the territorial impacts of natural hazards: The contribution of economic and public-private cooperation instruments. In T. Aven, & J. E. Vinnem (Eds.), *Risk, Reliability and Societal Safety, Vols 1-3: Vol 1: Specialisation Topics; Vol 2: Thematic Topics; Vol 3: Applications Topics* (pp. 603–611). Taylor & Francis Ltd.

Pesaro, G. (2015). Economie della resilienza per lo sviluppo dei sistemi locali,. In Quaderno N.21—Resilienza tra territorio e comunità—Approcci, strategie, temi e casiIn A. Colucci, & P. Cottino (Eds.), Fondazione Cariplo.

Ravi, R. V. (2019). *Partnership Among Grassroots Organizations*. MJP Publisher.

Rotmans, J., Kemp, R., & van Asselt, M. (2001). More evolution than revolution: Transition management in public policy. *Foresight (Los Angeles, Calif.)*, *3*(1), 15–31. Available from https://doi.org/10.1108/14636680110803003.

The tools of government: A guide to the new governanceIn L. M. Salamon, & O. V. Elliott (Eds.), Oxford University Press.

Voorberg, W. H., Bekkers, V. J. J. M., & Tummers, L. G. (2015). A systematic review of co-creation and co-production: embarking on the social innovation journey. *Public Management Review*, *17*(9), 1333–1357. Available from https://doi.org/10.1080/14719037.2014.930505.

PART III

Exploring resilience practices

CHAPTER 9

Challenging resilience of the agroecosystems in the Agro Pontino: focusing on ecosystem services and land stewardship in agri-environmental governance and planning

Federica Benelli[1], Mara Cossu[1], Eliot Laniado[1], Carlotta Sigismondi[1], Stefano Magaudda[2] and Serena Muccitelli[2]
[1]Poliedra—Politecnico di Milano, Italy
[2]U-Space srl, Italy

9.1 Challenging resilience in an intensive agricultural environment

The Pontine Plain is a reclamation environment that occupies a large part of the Province of Latina, an intensive and highly man-made agricultural environment in central Italy. The studies for the definition of the Provincial Ecological Network (Giunti et al., 2009) highlight three main conservation areas, identifying their priorities and needs. The humid (residual reclamation areas) and agricultural areas, which characterize the plain, appear to be deeply compromised in terms of habitat quality and biodiversity, both in terms of animal and plant species. Most critical issues relate to the loss of habitats necessary for the survival of the species, the increasing isolation of natural areas, and the reduction of the quality of natural and seminatural areas. The banks and windbreaks, recognized as the backbones of the network in these environments, show strong signs of suffering due to inefficiency or lack of management. Land consumption linked to increasing urbanization and the intensification of agricultural production, in particular greenhouse cultivation, represent the main pressures in this context, undermining ecological continuity.

[ECO]systems of Resilience Practices
DOI: https://doi.org/10.1016/B978-0-12-819198-9.00007-7

9.1.1 The Agro Pontino ecological network approach

In the last 20 years, some key players in the Pontine area have started a slow but constant territorial recovery and refunctionalization by exploiting the enormous potential for connection and regeneration of the green and blue linear systems of the Agro Pontino. The main actor of this environmental recovery and redevelopment path was the Province of Latina. In accordance with the Water Framework Directive (60/2000 CE), it launched several studies and monitoring activities on the territory and on its environmental components and challenges. Such commitment led to the adoption in 2010 of the Territorial Coordination Plan (Piano Territoriale Generale Provinciale - PTGP) as well as of the Provincial Ecological Network as an integral part of the PTGP. The follow-up of the ecological network represents a fundamental step toward the conservation of the Pontine habitats and the improvement of the quality and functionality of its ecosystems. Specific strategies are identified for agricultural environments, aimed at restoring continuity and reducing polluting loads.

Within those years a provincial wide-area planning approach based on environmental regeneration was designed—also through the contribution of *EU Life Programme*-funded projects—based on the cooperation of key stakeholders and backboned by concrete demonstrative actions. Such approach brought to evidence its potential in achieving considerable results in terms of raised awareness of institutions and citizens, as well as on the effectiveness of regulatory mechanisms for environmental planning and regeneration.

The approach was primarily shaped within the *Rewetland* project "Widespread introduction of constructed wetlands for a wastewater treatment of Agro Pontino" (LIFE + 08 ENV/IT/000406, http://www.rewetland.eu). It allowed the Province of Latina to approve the "Plan for the environmental regeneration of Agro Pontino." The project also implemented several demonstration actions, based on phytodepuration as a mean for restoring the quality of waters in the plain, contributing to biodiversity.

The Province of Latina further boosted the approach focusing on the implementation of the territorial ecological network by engaging the most "pollutant" and powerful actors in the Plain: the farmers. In intensive agricultural environments the effectiveness of environmental regeneration mainly depends on their behavior as well as on the potential transformation of agricultural practices.

The contribution of green infrastructures to the construction and functionality of ecological networks is essential and widely recognized in the literature as well as in community guidance and planning documents. The communication from the commission "Green Infrastructures—Strengthening Natural Capital in Europe" (European Commission, 2013a) and the accompanying working document (European Commission, 2013b) highlight how essential components of the infrastructures are not to be limited to the physical elements, although fundamental. Rather, they must expand to include a set of intangible elements that encompass initiatives aimed at preserving, improving, or restoring nature and ecological processes, integrated approaches to spatial planning as well as methods and techniques to activate the necessary investments to carry them out. In the Pontine context the proper use of greening, introduced by the Community Agricultural Policy (European Commission, 2011), as well as a correct use of the agri-environment means of the *II Pillar* can become a key reference for driving public and private action. The areas to be included in the provincial ecological network can in fact be compensated by funding the interventions under the next national strategic planning document (2021–27), still under definition. Company areas, particularly those devoted to cross-compliance and greening, can also contribute to the ecological network, for instance, through the Ecological Focus Areas, which have the greatest potential in terms of restoring natural biodiversity. An important coordination effort between different bodies and resources is required to activate complementary and synergistic measures and to ensure a public governance for connecting expected measures.

9.1.2 Combining initiatives and building a resilience-based governance for the agroecosystems

Starting from this assumption, the Province of Latina activated in time a series of initiatives targeted to the *material component* of the ecological network, particularly establishing protection regimes on *core areas* (SIC IT6040003 Gricilli Lakes; Natural Monument "Fosso Brivolco and limestone surfaces with dinosaur footprints") and connecting them through river regeneration. At the same time, following this twofold approach, it started working harder on the "intangible component" of the network.

In 2018 it obtained funding for a new LIFE proposal developed in collaboration with local actors, technical–scientific partners and a Maltese institution. The *Greenchange* project "Green Infrastructures for increasing biodiversity in Agro Pontino and Maltese rural areas" (LIFE17 NAT/IT/000619, http://lifegreenchange.eu/it/, hereinafter Greenchange) aims to preserve biodiversity and enhance ecological values of the agroecosystems. To test and improve the provincial wide-area planning approach, in the Agro Pontino, Greenchange combines demonstrative actions aimed at the construction of green infrastructures with the definition of decision-making processes and governance mechanisms based on the evaluation of ecosystem services (ESs).

After more than two decades since the beginning of the environmental recovery process of the Agro Pontino, which made available over 10 million euros, the first concrete results are increasingly visible. The wide-area approach, based on strong public direction and the cooperation of key actors, has revealed to be essential in ensuring the effectiveness of local governance practices. The provincial administration had the value of developing a strategy for the recovery and enhancement of the territory by considering the Pontine Plain (about 700 km^2) and the network of reclamation canals (about 2200 km) as a single target for planning and action. The Province of Latina also had the ability to overcome the contrasts between sectoral and institutional areas of competence, by following a collaborative governance approach, leading to a strategic, multilevel, and resilience-based vision that combines integrated management of water resources, protection of biodiversity, mitigation of climate change, ecosystems improvement, reduction of land consumption, and sustainable practices in rural areas.

In this view, concrete multifunctional interventions did address functional rehabilitation of urban, peri-urban, and rural areas, contributing to the achievement of objectives concerning depollution (DEP), landscape (LAND), soil (SOIL) and biodiversity (BIO) protection, sustainable agriculture (AGRI), recreation (REC) (see Table 9.1 and Fig. 9.1).

9.2 Engaging actors for biodiversity in agriculture: the pact for biodiversity

Together with the consistent amount of public resources provided, experience in the Agro Pontino reveals the critical role played by farmers in creating and maintaining the ecological network, properly considering their poor responsiveness in taking advantage of climate and environment funding measures (see Fig. 9.2). Greenchange argues the need for their

Table 9.1 List of the multifunctional interventions implemented in the Agro Pontino and the related objectives.

		DEP	BIO	SOIL	LAND	AGRI	REC
PAN	Pantanello/Ninfa wetland	○	•	○	•	○	•
RW1	Fogliano filtering ecosystem	•	•	○	•	○	•
RW2	Marina di Latina phytodepuration	•	•	○	○	○	•
RW3	Buffer strips along reclamation channels	•	•	○	○	•	○
RW4	Demonstrative project in Casale del Giglio farm	•	•	○	○	•	○
PLUS	Phytodepuration at wastewater treatment plant	•	○	○	○	○	○
MON	Cava di Monticchio Natural Monument	○	•	○	•	○	•
QA1	Sermoneta urban park	•	•	•	•	○	•
QA2	Pontinia linear park	•	•	•	•	○	•
QA3	Mole Muti phytodepuration	•	•	○	•	○	•
QA4	Latina urban phytodepuration	•	•	○	○	○	○
QA5	Aprilia urban phytodepuration	•	•	○	○	○	○
GC1	Gricilli wetland	•	•	•	•	•	•
GC2	Environmental restoration at Lago Pani	○	•	○	•	○	•
GC3	Improvement in Pantanello wetland	○	•	○	•	○	•
GC4	Demonstrative project in G. Caetani farm	•	•	•	•	•	•
GC5	Restoration of Cicerchia channel—Ganci farm	•	•	○	○	•	•
GC6	Restoration of Allacciante channel—Agrilatina farm	•	•	○	○	•	○
GC7	Demonstrative project in Roana livestock farm	•	•	○	○	•	○
GC8	Improvement of windbreaker strips	•	•	○	•	•	○

active engagement to create, maintain, and enhance linear ecological connections (windbreaks and banks, canals, dry stone walls).

To these combined aims, Greenchange project promotes farm planning and management methods based on the assessment and the enhancement of ESs provided by the agroecosystems. It builds on the remunerability that ecological areas can have in the context of the new Common Agricultural Policy (CAP) and of the Rural Development Programme (RDP) of Lazio Region, which remain the main funding support for farmers. Based on such potential, it promotes farmers'

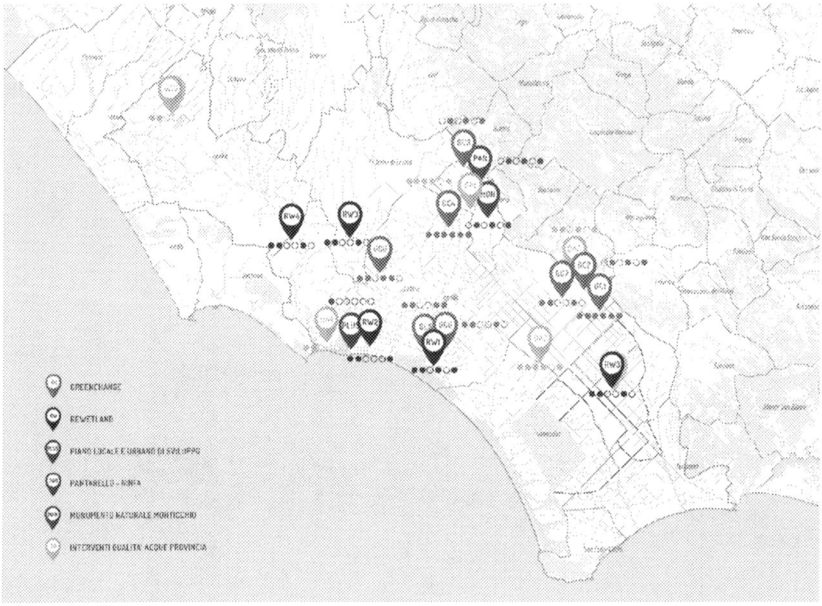

Figure 9.1 Localization of the multifunctional interventions in the Agro Pontino.

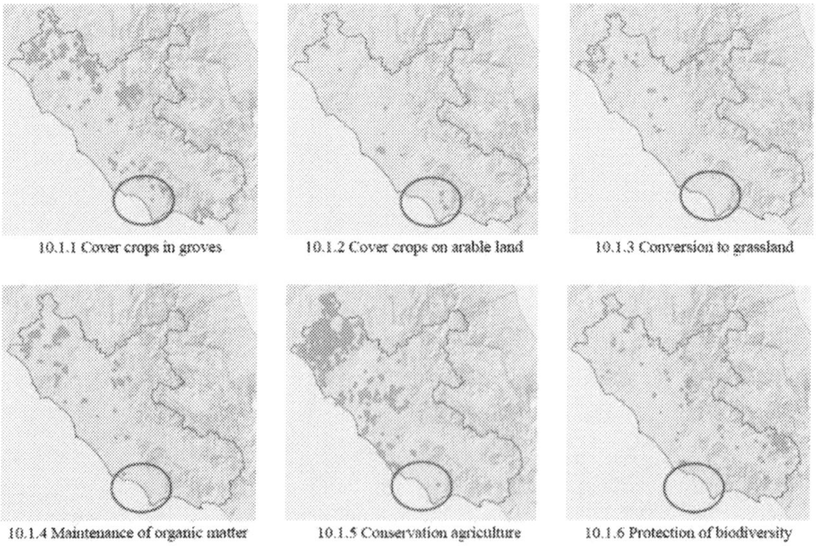

Figure 9.2 Distribution of agro-environment measures of the Rural Development Programme 2014/2020 of Lazio Region (Pontine Plain in circle) (COGEA, 2019).

engagement particularly by spreading the importance and remunerability of areas of ecological interest within agricultural holdings, which can contribute to the design of the ecological network in the Agro Pontino, while proposing defining priority criteria for the RDP based on farmers' potential as "land stewards."

In supporting the rebalancing of the relationship among funding schemes rules and procedures and potential beneficiaries, it addresses main factors recognized as the main obstacles in the achievement of environmental protection objectives in the Agro Pontino:

- lack of awareness of farmers on the environmental impacts related to agronomic practices (both intensive and extensive) and on their role in preserving agricultural areas of ecological importance;
- little involvement of farmers in conservation measures and in the implementation of management plans for large areas or Natura 2000 areas;
- nonattractiveness of RDP incentives (lack of knowledge, excessive bureaucracy) and need for territorial-targeted support;
- lack of territorial and/or "beyond-farm" perspective guidelines and governance, to fully implement and benefit from the greening perspective of the next CAP to concretely support ecological networks in intensive rural environments.

Combining the awareness about production and cropping services provided by agroecosystems and huge potential *green* funding from CAP and RDP, the project also boosts networking and collaboration among farmers, to both ensure the continuity of interventions for conservation purposes and to counteract the abandonment of funds. In this frame the role of institutions and farmers' associations is key to provide the proper transformational and collaborative environment.

Such collaborative governance has been created within a so-called Pact for Biodiversity in the Agro Pontino, promoted in 2020 by the Provincia di Latina, and signed by Lazio Region, the Ministry of Agricultural, Food and Forestry Policies, the Water Reclamation Authority of Latina, the Management Authority of the Circeo National Park, the local technical school for agriculture and the environment, the three main farmers' unions of the Provincia (Confagricoltura, Coldiretti, CIA), and other partners.

A pivotal role in the new governance system will be played by land stewardship (LS) (Sabaté et al., 2013), an approach promoting the voluntary assumption of coresponsibility for the sustainable management of natural resources in rural areas, which is currently being used in the testing and implementation phase of the project. More specifically, LS agreements (LSA) are a constituent element of Pact for defining the farmers' commitments for the implementation of the ecological network in the Agro Pontino.

The Pact together with related beyond-farm guidelines, strategic vision, and operational funding criteria works for the engagement of all stakeholders both at the administrative and at the holding level. The Pact gathers all institutions dealing with the management and implementation of agricultural policies and land management, willing to harmonize and simplify rules and procedures, while enhancing their effectiveness. To ensure the enlargement of the spectrum of engaged actors, the Pact sets each actor's rule and responsibility. Continuous training, advisory, and awareness-raising activities are also taking place to stimulate the active involvement of farmers and their associations in the mitigation of the environmental impacts of farming and in the enhancement of the services provided by agroecosystems.

The broader aim of the Pact is to engage with competing multiple rationalities and with conflicts arising from the different values that all actors represent and bring into the decision-making and implementation territorial arena. It follows a tentative and collective ongoing process, led by the so-called adaptive management (Holling, 1978). Intended in a wide sense, it allows establishing an iterative territorial learning process enriching knowledge and delivering short-term outputs for management.

In this sense, it stands as an open and permanent working table, that institutions and farmers' associations in the Agro Pontino are called to transform into a learning environment for associated farmers, also regarding the potential and mutual role that the functionality of ecosystems plays in the management and productivity of the rural territory. It then focuses on education, information on the value of agroecosystems for biodiversity conservation as well as for crop productivity and functionality, accountability, and calibration of funding through simplification and "intelligent" reward systems.

The Pact is progressively widening the protected natural areas designated as *core* elements of the network, promoting concrete demonstrative actions on windbreak strips and buffer strips along the reclamation canals' banks, and assessing such actions to test the effectiveness of LSA. In this perspective the creation of the ecological network is interpreted not only as an essential element for the conservation of habitats and species, but also as an important factor for the rebalancing and productivity of agricultural soils, for recharging groundwater, for contrasting the local effects that climatic change is displaying also on crops. Following this argument, the Pact also works for simplifying the authorization and assessment procedures linked to investments in agricultural holdings through recognizing the undertakings signed within the LSA.

The Pact is setting operational instruments in the closure of the current CAP cycle 2014/2020 and in the wake of the reform of the CAP launched in 2017 (European Commission, 2017, 713; European Commission, 2019, 580 final), still pending final approval due to Covid-19 pandemic. This reform will maintain the main approach to support agriculture, but will differentiate mechanisms, pushing increased attention to climate and environment issues. It will imply a substantial strengthening of basic conditionality, strengthening the commitment to greening through the transition from diversification to crop rotation. Furthermore, additional support is envisaged for farmers who engage in practices that are advantageous for the climate and the environment, in synergy with the agri-environmental commitments envisaged in the RDPs, through an innovative system of conditionality known as "Eco-schemes," the modalities of which are being fully defined. In this vision, CAP 2021/2027 will support the ability of agriculture to contribute to the green infrastructure of the territory by also providing productive rural spaces and promoting agricultural practices that strengthen the functionality of agroecosystems. The Pact is working to anticipate such approach and prepare farmers and territorial institutions to fully benefit from it. A huge effort is being spent to modulate the subsidizing systems and rewarding criteria for the agri-environmental measures of the RDP of the 2021–27 programming period under preparation, by continuous and inclusive negotiating process with the regional authority.

9.2.1 Ecosystem services assessment supporting decision-making

To properly set operational instruments and actors' attitude, the Pact relies on a contextual analysis able to enlighten the dynamics influencing the capacity of the agroecosystems to provide expected services, as well as to identify interventions and practices that can support such capacity, while providing positive effects on farm management (cost savings, resource, and environmental risk management).

The Agro Pontino has been analyzed o identify main "ecosystems" at territorial (1:25.000) and at action (1:5.000) scale, through progressive aggregation of land use classes. The analysis stressed the role of residual natural elements within the "rural pattern" (European Parliament, 2000). Such residuals, properly combined with environmental restoration interventions within the intensive rural land, can evolve as widespread patterns on a landscape scale, capable of favoring ecological processes and keeping populations and biological communities vital over time (Giunti et al., 2009). The "ecosystem map" of the Agro Pontino selected five natural and seminatural ecosystems, two humid ecosystems, and five agroecosystems, classified on the level of water–chemical–mechanical input required (see Fig. 9.3).

After mapping the ecosystems, relevant ESs in the Agro Pontino have been selected, under CICES 5.1 classification (European Environment Agency, 2018). The selection was provided following three criteria: relevance toward the environmental compartment, potential relationship with demonstrative actions, capacity to be described by available and easily updatable indicators. The selected ESs are:

- cultivated terrestrial plants grown for nutritional (1.1.1.1);
- maintaining nursery populations and habitats (2.2.2.3);

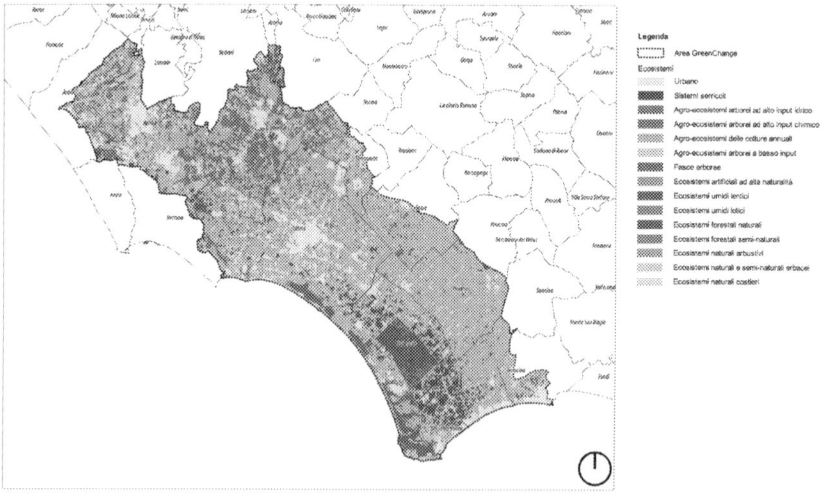

Figure 9.3 Map of the Agro Pontino ecosystems. *From Life-Greenchange Deliverable A1 Mapping and assessment of the ecosystem and ecosystems' services.*

- regulation of the chemical condition of freshwaters by living processes (2.2.5.1), focused on the phyto-purification capacity offered by the riparian systems;
- regulation of chemical composition of atmosphere and oceans (2.2.6.1), focused on carbon storage;
- direct, in situ and outdoor interactions with living systems that depend on presence in the environmental setting (3.1), focused on the opportunities to enjoy the rural landscape and its values.

ESs have been assessed through quantitative and qualitative methods. Quantitative analysis was carried out by using dedicated and wide-tested software (Neugarten et al., 2018), INVEST and Tessa model, to guarantee comparability. To provide adequate insights on the interactions among the context, the services provided by agroecosystems and all initiatives and dynamics acting on the Agro Pontino, a qualitative analysis has been carried out, using the resilience and vulnerability analysis (Cossu et al., 2015) (Vandergert et al., 2016).

Resilience, defined here as the ability of a territorial system to absorb shocks and progressive changes, reacting and evolving to maintain identity, structural functions, capacity of self-organization (Project CARE report, adapted from the IPCC, 2007—http://www.erasmus-care.eu), has been used as a key for a holistic assessment of territorial dynamics in and for planning and designing integrated and multidimensional initiatives, such as the Pact for Biodiversity. Setting a dynamic knowledge basis, the decision-making system within the Pact is thought to be dynamic and adaptive itself, capable of reflecting on territorial effects displayed by its means. The analysis is based on landscape and environmental systems, defined as the main elements for delivering complex descriptions of current territorial dynamics and describing the characteristic qualities and responsiveness of each area. Such responsiveness is determined by pressures and factors that are intrinsically part of the systems and can both qualify them (increasing their resilience) or compromise them (worsening their vulnerability). For identifying the territorial dynamics, statistical analysis, comparison with actors, and analysis of the planning and policy framework have been used.

Fig. 9.4 summarizes in the form of a conceptual map—as a way of communicating complexity to improve knowledge (Borrego et al., 2009; Sibilia, 2017)—those that have been identified as the main dynamics that can compromise or improve the ability of the Agro Pontino to provide some ESs. The dynamics of vulnerability (left) and resilience (right) are

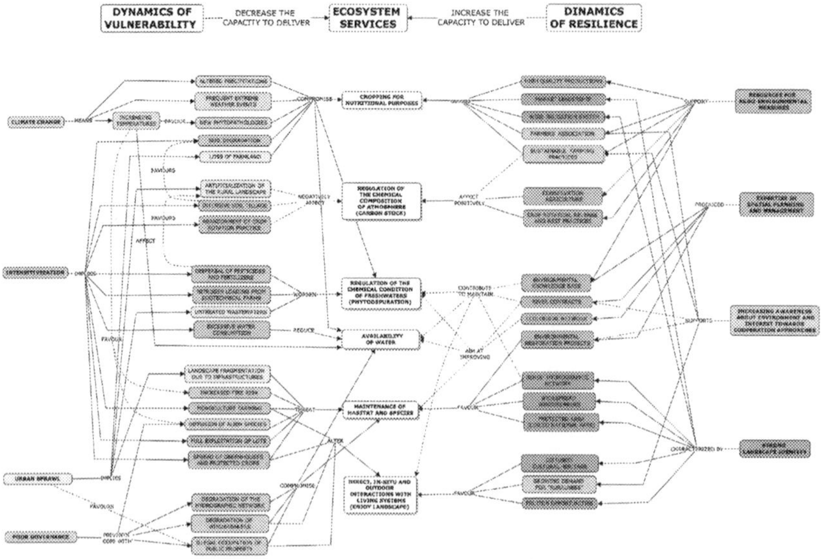

Figure 9.4 Resilience and vulnerability analysis applied to the ecosystem services in Agro Pontino as a concept map.

identified grouped by "macrodynamics" and distinguished by color. At the center are the five ESs subject to in-depth analysis (V/R concept map).

Following the identified dynamics, the Pact selected its operational objectives to shape the action plan of the Pact in the long term: protecting residual natural elements, reducing the chemical inputs, optimizing water consumption, tackling soil degradation, acknowledging and improving agricultural biodiversity, maintaining/restoring linear landscape elements, maximizing carbon stock, maintaining farming profitability and diversifying farmers' income sources, limiting the uncontrolled spread of greenhouses.

Three fields of action have been defined to pursue such objectives in the medium term, during the Greenchange implementation—producing the so-called project deliverables—and beyond, in the "after-life."

The first field of action concerns building up a governance system: Greenchange planned to issue two different guidelines, both targeted to the regional policy makers. The first focuses on the opportunities to ease the authorization procedures and environmental requirements for green infrastructure works within farms. The second proposes selection and localization criteria to apply to the forthcoming rural development funding instruments to support agri-environmental measures in the intensive farmland.

The second field of action concerns tools to support farmers as land stewards. To the purpose, Greenchange is sponsoring an information desk at the local office of one of the Italian main farmers' associations, Confagricoltura. Through the desk, a Geographical Information System tool is being tested, aiming to advise farmers on green infrastructures implementation and funding potential, based on mapping and evaluation of Agroecosystem Services. Several technical documents were also produced, such as the catalog of agri-environmental interventions, the catalog of the impacts caused by agricultural practices, the guidelines for the management of irrigation channels and windbreak strips.

The third field of action concerns the promotion of LSA. In the context of the demonstrative projects financed in the Agro Pontino within the Greenchange project, an innovative contractual form of loan for use has been tested with the partner farms that host the interventions (Fig. 9.5). Greenchange has undertaken to sign 50 agreements of this type 5 years after the end of the project .

9.2.2 Supporting the "guardian farmers": the land stewardship agreements

As already mentioned at the heart of the LS, there are voluntary agreements between owners, land users, and LS organizations. The LS develops an experimental tool that working on multiple levels is particularly useful when the goal is the sustainable management of natural resources, rather

Figure 9.5 Drone view of the Ganci winegrower holding that hosts one of the Greenchange demonstrative projects.

than their absolute protection, but also the protection of open spaces and the most fragile natural areas from the development pressures, especially in those areas where planning and control are weak. Given this definition, sustainable agriculture and connected ESs can play an important role in the protection of the territory, contributing to the development of positive externalities in the environment and in the sphere of public goods. LS can therefore play an important role both within the Natura 2000 network of protected areas and outside. The latter, which falls into public and private land, urban or rural, and is not subject to specific protection regimes, is even more important in improving the ecological connectivity of the territories, namely, the network of green infrastructures.

A peculiar character of the LS is the recognition and enhancement of the role of private actors in the conservation of nature, which in turn activates a subsidiarity process that widens the sphere of protection of the landscape and ecosystems within a wider participation network. The implementation of nature conservation actions therefore contributes to the sharing of the values of the territory, both in peri-urban areas and in rural contexts.

As it has been analyzed by the LandLife project (LIFE10 INF/ES/000540), LS is present in slightly different forms in all European countries, in all cases responding to the need, shared at EU level, of a renovated governance capable of redistributing benefits and responsibilities within actors active in a particular field (Sabaté et al., 2013). From the implementation point of view, LS offers innovative and quality solutions for farmers and owners useful to strengthen their production, the services they offer, and not least their environmental engagement. The innovative character of these tools lies in the fact that they enable complex forms of coproduction of environmental goods with a higher relational and cognitive content, which would not be possible to obtain through the sole commitment of the State or the Market (Donolo, 2005). The stakeholders involved in the LS can be grouped as *actors*, landowners and land users that voluntary commit in the conservation of nature (farmers, hunters, fishermen, foresters); *facilitators*, LS organizations that subscribe the agreements with the actors, monitor the conservation actions, and develop the guidelines for the sustainable management of nature (nongovernmental organizations, local administrations); *enablers*, governing bodies (local, regional, central, or EU) who create the basis for cooperation within the first two groups; *general public*, beneficiary civil society.

The integration of LS into the Pact for Biodiversity of Agro Pontino offers one more layer to the complex agri-environemental governance chain that the Pact is building for the implementation of green infrastructures allowing to foster a collaborative approach for the management of natural resources. By enabling the development of synergic and coordinated actions between farms, local authorities, and the other actors engaged in regional governance, LS appears to be a very effective approach to guarantee biodiversity conservation and rural landscape valorization (Chiodo & Vanni, 2014). In the specific case of Agro Pontino, three actors are key: farmers—involved in production, public and managing authorities—responsible of defining the strategy and bringing their specific knowhow, and the regional administration that promoted ad hoc measures and funding (Prager et al., 2012).

To engage farms in the sustainable maintenance and management of the secondary network of canal banks and windbreaks, specific models of LSA have been developed in the framework of Greenchange. They aim to formalize farmers' commitments to the conservation and restoration of nature on their property and in state-owned areas, the management of which has been transferred to them though loan for use. This mechanism allows to enhance access to CAP funding, being based on the possibility for farmers to fulfill the greening requirements on state-owned GI. Through loan-for-use contracts with the region, that is the owner and the body in charge of managing windbreaks and canal banks, such areas become part of the farm dossier and are eligible for funding.

It is possible to highlight that in the target context, the agreements would bring benefits to all the subjects involved: from farms to the region. The benefit farmers receive is twofold: besides CAP direct payments, they earn the possibility of devoting the transferred areas (outside their productive land), which contribute to expanding the respective farm area, to comply with the criteria imposed by the CAP regarding greening and to have easier access to agri-environmental measures. The Lazio Region, by transferring the responsibility of the secondary network's maintenance to agricultural companies, can allocate the economic resources saved to the more substantial maintenance and priority requalification of the primary network. Finally, the Province of Latina accomplishes to promote the implementation of its planning and programming tools and to allocate resources on governance and monitoring activities. Within the benefits that it is expected to get from the implementation of LS, it is important to also highlight the encouragement of collective knowledge production, thanks to the actors facilitating the farmers' learning process, which in turns aims at the promotion and creation of autonomy and cooperation in farmers.

One last pivotal issue for the successful implementation of the agreements concerns the complexity of the process that leads to the actual drafting of the LSA. These are in fact the result of a process of consultation and negotiation where several pragmatic nodes emerge, such as bureaucratic aspects, the specific needs of the territory and of the engaged actors, not only in productive and functional terms, but also in terms of social recognition.

9.3 A perfect recipe, what's missing?

Greenchange experience highlights how ecosystems management at territorial level is quite committing. Mainly, such complexity is due to stakeholders' engagement and management, including landowners and dealers, multilevel institutions and competences, decision makers, experts, and the general public. The stakeholders' involvement since the beginning of the decision process increases policy acceptance and eases negotiations (Prager et al., 2012).

In this view a key for future CAP as well as for planning policy in agro-territories is the active engagement of all stakeholders. New programs should incorporate local knowledge, in addition to expert views, to build trust, and to encourage acceptance of the schemes.

The Pact for the Biodiversity becomes in this frame a powerful and incremental engagement platform. If properly used, it can support the acceptance of planning and funding measures to boost environmental rehabilitation and ecological reconnection of the Plain. It is to be intended as an open and adaptive governance mean (Gabaglio & Silvestri, 2014), based on the simplification of administrative formalities and permits, avoiding banalization of environmental duties, as well as on optimizing the use of funding through exploiting synergies. It should generate community empowerment and increase local awareness while strengthening problem-solving attitudes.

The main weakness is being recognized in policy mandate and in the lack of collaborative attitude by institutions (vertical governance). The relevant overlapping and fragmentation of competencies in land management make it very difficult to turn to shared decisions and management schemes. It implies a long-term vision and approach which conflicts with the dramatically short-term perspective of policy making.

Furthermore, current agri-environmental schemes that compensate farmers for managing land in an environmentally friendly way are typically targeted at the local level of management. This often results in uncoordinated actions across the landscape level. Also, the voluntariness of collaboration can result in a lack of commitments from all actors involved. A collaborative system based on respective duties, as the LSA can provide, can help avoiding mistrust among groups of stakeholders. This also applies to the monitoring and assessment governance, to share in advance appropriate monitoring schemes and appropriate governance to ensure the collection of needed data and their elaboration at all relevant scales of management.

Compliance with the scheme will be increased if land managers participate to the availability of data to implement selected indicators. Moreover, a direct linkage among ecological efforts and funding is to be recognized by funding criteria, to stimulate farmers' commitment.

The potential relationship between the Pact, the LSA, and the collaborative approach being integrated into the new CAP could significantly improve bottom-up effectiveness of the measures and the achievement of the objectives set by the Pact itself.

Sufficient funding and adequate priority are to be assigned to collaborative programs and interventions. Moreover, to ensure farmers join the scheme, payment levels need to be high enough to compete with other opportunities for land use and other sources of income available to farmers, in intensive farmland as well as in inner areas.

The Pact can be the environment where to deal with all such complex and interconnected issues, if proper political mandate and institutional collaboration are provided. Close relationships created across all management levels will enhance the ability of ecosystems to provide essential services, benefiting from a stable, permanent and multistakeholder collaborative environment.

References

Borrego, M., Newswander, C. B., McNair, L. D., McGinnis, S., & Paretti, M. C. (2009). Using concept maps to assess interdisciplinary integration of green engineering knowledge. *Advances in Engineering Education*, *1*, 3.

Chiodo, E., & Vanni, F. (2014). La gestione collettiva delle misure agro-ambientali: oltre le esperienze pilota? *Agriregionieuropa anno*, *10*(36). Available from https://agriregionieuropa.univpm.it/it. (March 2021).

COGEA (2019). Servizio di valutazione indipendente del Programma di Sviluppo Rurale 2014–2020 del Lazio, Rapporto di valutazione intermedia. Available from http://lazioeuropa.it/files/190606/rapporto_val_intermedia_maggio_2019_def.pdf (March 2021).

Cossu M., Pezzoli S., Vaghi, S. Comprendere, valutare e monitorare le strategie di resilienza, in Resilienza tra territorio e comunità – Approcci, strategie, temi e casi, a cura di A. Colucci A. P. Cottino. Collana "Quaderni dell'Osservatorio" n. 21 Anno 2015 Fondazione Cariplo.

Donolo, C. (2005). Dalle politiche pubbliche alle pratiche sociali nella produzione di beni pubblici? Osservazioni su una nuova generazione di "policies,". *Stato e mercato*, *73*(1), 33–65.

European Commission (2011). Common agricultural policy towards 2020. Assessment of alternative policy options. SEC(2011) 1153 final/2. Brussels.

European Commission (2013a). Enhancing Europe's Natural Capital. COM/2013/0249 final. Brussels.

European Commission (2013b). Technical information on Green Infrastructure (GI) - Commission Staff Working Document. SWD (2013) 155 final. Brussels.

European Commission (2017). Future of food and farming. COM/2017/713 final. Brussels.

European Commission (2019). Proposal for a REGULATION OF THE EUROPEAN PARLIAMENT AND OF THE COUNCIL amending Regulation (EU) No 1306/2013 as regards financial discipline as from financial year 2021 and Regulation (EU) No 1307/2013 as regards flexibility between pillars in respect of calendar year 2020. COM/2019/580 final. Brussels.

European Environment Agency (2018). Common International Classification of Ecosystem Services (CICES) Spreadsheet Version 5.1. https://cices.eu/ (March 2021).

European Parliament (2000). Resolution on the application of the Convention Concerning the Protection of the World Cultural and Natural Heritage in the Member States of the European Union (2000/2036(INI)).

Gabaglio, E., & Silvestri, F. (2014). I contratti di fiume in Lombardia: un modello di governance per le politiche territoriali. *Agriregionieuropa anno*, *10*(37). Available from https://agriregionieuropa.univpm.it/it. (March 2021).

Giunti M., Piazzi A., Forte A., a cura di (2009). *La rete ecologica della Provincia di Latina*—Relazione Tecnica Conclusiva.

Holling, C. S. (1978). *Adaptive environmental assessment and management*. John Wiley & Sons.

Neugarten R. A., Langhammer P. F., Osipova E., Bagstad K. J., Bhagabati N., Butchart S., Dudley N., et al. (2018). Tools for measuring, modelling, and valuing ecosystem services IUCN.

Prager, K., Reed, M., & Scott, A. (2012). Encouraging collaboration for the provision of ecosystem services at a landscape scale – Rethinking agri-environmental payments. *Land use Policy*, *29*, 244–249.

Sabaté, X., Basora, X., O'Neill, C., & Mitchell, B. (2013). Caring together for nature. Manual on land stewardship as a tool to promote social involvement with the natural environment in Europe. *LandLife documents*, *Volume online*, First edition.

Sibilia, M. (2017). A meaningful mapping approach for the complex design. *International Journal of Design Sciences and Technology*, *23*(2), 41–78.

Vandergert, P., Collier, M., Kampelmann, S., & Newport, D. (2016). Blending adaptive governance and institutional theory to explore urban resilience and sustainability strategies in the Rome metropolitan area, Italy. *International Journal of Urban Sustainable Development*, *8*(2), 126–143.

CHAPTER 10

Toward a framework for assessing urban resilience. Application to a case study in Lyon (France)

Nicolas Beaurez[1], Lise Bourdeau-Lepage[2], Marie-Laure Couderc-Papaix[1], Nicolas Furmanek[1] and Sarah Talandier Lespinasse[1]
[1]Cerema, France
[2]CNRS UMR EVS, Université de Lyon, Lyon, France

10.1 Introduction

Resilience is a polysemous concept that is used in different disciplines such as psychology, materials physics, ecology, geography, and economics (Bruijne et al., 2010; Cutter et al., 2008; Hernandez, 2009). Several definitions coexist and are sometimes different or even contradictory. Resilience can be considered the ability to function in degraded mode (Pigeon & Rebotier, 2016). It can be seen as the ability to withstand a shock without damage or to absorb and recover from a shock (Timmerman, 1981). It is also defined as the ability to adapt to, or resist a hazard. In the field of natural and technological risks, resilience contributes to disaster prevention and reduction and may reflect the aim of putting an area or development operation back to normal or bringing it into equilibrium. This is because an area or development operation is potentially vulnerable and may be subject to hazards, that is, the possibility of an extreme event occurring that would cause damage and malfunctions. They are qualified as resilient if, whenever this event occurs, they return to their initial situation or rebuild themselves as they were or differently. Resilience is therefore generally understood as the ability to absorb shocks and maintain integrity, through a return to a state of equilibrium or a process of reconstruction (Holling, 2001). Resilience procures the means to react, absorb shocks, renew, and reorganize.

However, an approach to resilience based mainly on shocks does not seem sufficient to address the challenges facing urban spaces. It is of

[ECO]systems of Resilience Practices
DOI: https://doi.org/10.1016/B978-0-12-819198-9.00015-6

fundamental importance to take into account chronic stresses such as drought, economic decline, the receding coastline, and dependencies such as the dependence of the economy of a mid-mountain ski resort on snowfall or the climate dependence of a farming activity. It is all the more important since these threats may be interdependent. Climate change may, for example, increase water stress in some regions or amplify certain extreme weather events. Where urban spaces are concerned, the example of Paris helps us to understand the complexity that a city faces when it seeks to become resilient and the variety and interdependence of the issues involved. In its resilience strategy, this city identifies six major issues: (1) socioeconomic and spatial inequalities; (2) terrorist risk and its economic and psychological impacts; (3) climate change and its impacts: heat waves, drought, violent storms, floods, stress placed on water resources, with impacts on urban heat zones, air pollution, fragility of transport infrastructure and urban services, excess mortality, especially of vulnerable people; (4) air pollution as chronic stress and its impacts on health, the economy, and biodiversity; (5) river disturbances (flooding and subsequent decline in water resources); and (6) local governance (understood as interdependence with the areas surrounding Paris, the suburbs, and rural areas for food or catchment areas for water management). These issues are not exhaustive and could also include nuclear risks, pandemics, financial and oil shocks, and massive strikes (Hatchuel et al., 2017). They correspond not only to shocks, such as terrorist risks and flooding, but also to stresses, such as heat waves, air pollution, inequalities, or even dependencies, such as the interdependence of areas for food or water resources.

In an urban setting, it therefore seems important to make a central feature of the design of resilience the ability to anticipate to reducing the consequences of a shock and stress and dependencies. Our way of conceiving resilience therefore emphasizes long-term action and stresses the logic of anticipation and adaptation or even transformation because it is indeed a question of developing *abilities*, *capacities*, or *capabilities* (Sen, 1985). Taking time into consideration allows us to embrace a wide range of potential threats: shocks, but also stresses and dependencies. We think of resilience in a systemic way because, in our opinion, technical, social, economic, and spatial factors have an impact on each other (Cerema, 2019).

Today, and in response to the 2010 UN campaign *Toward Resilient Cities*, city practitioners are seeking to improve the resilience potential of the areas or development operations for which they are responsible, which means assessing this potential.

In economics, cost–benefit analysis (CBA) is used to assess the benefits and costs of implementing resilience measures. However, this analysis is difficult to carry out, particularly since the measures are numerous and systemic. A number of them are even what are known as transaction costs. The benefits are typically assessed in terms of the costs of the damage avoided. The difficulty is therefore to monetize the various positive and negative external factors. However, in the absence of a reference value (such as in transport where the State decides the monetary value of the deaths avoided or the time saved), the exercise is a difficult one.

In addition, the development operations to be assessed have, above all, urban functions. They are not designed only to combat a hazard or stress, as in the case of a dike that may have cobenefits but the primary function of which is related to managing a hazard. It is therefore a question of designing urban developments that perform their own function and are resilient. Assessing them by using CBA is tricky, especially when it is known that: even if this CBA "assessment tool is becoming popular, the methods, elements and indicators needed to adequately identify the whole of the costs and benefits to be considered are still under development" (Pesaro, 2018; 4).

So while the theoretical basis is not yet stabilized, practitioners need a method that can be used ex ante and ex post to assess the degree of resilience of a development and consider improving it. The assessment of resilience must therefore be able to determine whether, and to what extent, a development or area is adapted to cope with shocks, stresses, and dependencies. In this context the development of a tool that helps urban planners and developers to design urban developments that can withstand or recover from shocks, stresses, and potentially harmful dependencies is highly desirable. Such a tool has been designed by Cerema, the French public expert body on ecological transition and regional cohesion. This assessment tool takes the form of an analysis table.

This chapter presents and tests this tool, designed as a table for analyzing the resilience of areas or developments. Known as the "Resilience Framework," it is easily applied across all urban areas and on all objects within the public space (roads, buildings, etc.): from towns to public spaces and any urban planning.

We will first present the "Resilience framework" tool designed by Cerema (Section 10.2). We will present the levers on which it is based (Section 10.2.1). We will then explain how we used the framework on a development operation, and the alternative methods that can be used (Section 10.2.2).

Next, we will report on how the tool was applied to a urban planning operation that has now been finalized, to illustrate the assessment possibilities that the tool can actually provide (Section 10.3). We will see that the development work examined was considered particularly resilient, first because of its ability to take into account the various stakeholders and to incorporate risks as of the design stage to reduce vulnerabilities and also because of its ability to anticipate (Section 10.3.1). We will highlight possible areas for improvement, in particular the ability to strengthen local identity and encourage a culture of risk (Section 10.3.2). We will conclude with considerations about the tool, the assessment of resilience, in general, and possible methods of using the tool.

10.2 A collective intelligence tool for territorial resilience: the resilience framework

The purpose of this section is to present the assessment tool developed, known as the resilience framework (Section 10.2.1) and the method used to administer it (Section 10.2.2).

10.2.1 Designing a tool for collective intelligence and analysis of the resilience of an area or development work

The objective is to design a tool that can be used to assess whether urban development or an area, taking into account its own specific issues, is able to continue to perform its urban functions in the event of shock, stress, and dependence and then return to functional equilibrium, or even adapt or transform itself. This tool must fulfill several functions. It must help planners and local authorities to make decisions when choosing between different projects for the same development and also allow the project to be assessed once it has been designed.

10.2.1.1 Empirical construction of the framework

The framework is based on six levers. The levers are mainly derived from work carried out and monographs produced by Cerema as part of the "Integrated analysis of territorial resilience" (*Analyse Intégrée de la Résilience des territoires*—AIRT) project led by the French commission for sustainable development of the Ecology Ministry (*Commissariat général au développent durable*—CGDD). They were heralded in the publication "*La résilience des territoires aux catastrophes* (The resilience of regions to disasters)." The levers have also been presented to, and developed with, many people working in the field during operational seminars (Cerema, 2016, 2017, 2019a,b). This feedback has led to the emergence of 6 levers and 18 subcomponents as a common denominator promoting resilience dynamics (see Fig. 10.1). This framework applies to all types of regions or public policies in a context of global resilience. It can also be used on more specific urban planning, across a district, for example, on specific themes (mobility, etc.) and within the context of one or more specific risks.

10.2.1.2 The six levers of resilience

These levers of the resilience framework define, so to speak, the main actions to be carried out across an area or within a project to move toward greater resilience.

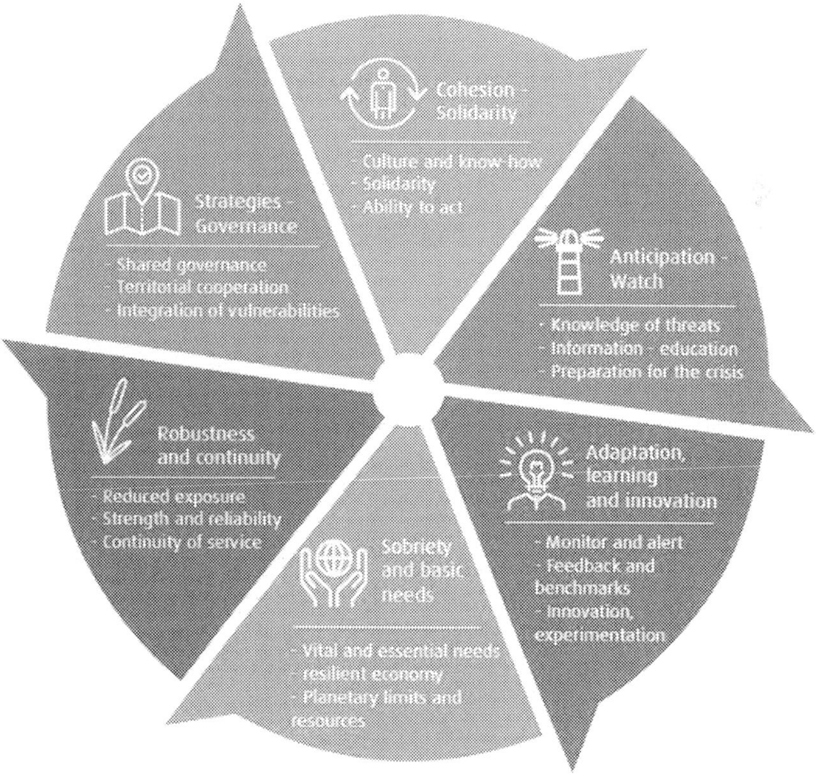

Figure 10.1 The six levers of resilience.

The first lever is the "integrated approach to the area." Regions are facing many risks, some of which are new. This lever makes it possible to cross-reference sectoral themes such as infrastructure, buildings, the environment, and energy. Several scales of time and area are taken into consideration. An integrated approach will take cobenefits into account. For example, nature-based solutions will help address urban heat zones, reduce runoff, promote biodiversity, and improve air quality. These solutions will also contribute to the well-being of users. Introducing actions with multiple benefits in different areas is a sign of resilience. Conversely, protecting against flooding by building a dike, as an isolated project, may have adverse effects by limiting the area of flood expansion, although floods fertilize agricultural land and promote the reproduction of certain fish species. Dikes make wetlands disappear and increase the speed of currents that contribute to erosion, not to mention the maintenance costs that these structures incur. As will have been understood, the difficulty is to assign a value to all the positive and negative externalities. Considering the area in its entirety, integrating risks, opting for cross-disciplinary and multibenefit approaches, and promoting cooperation between regions is the solution adopted.

The second pillar is entitled "Stakeholder diversity and cooperation" (inclusive approaches). Developing a regional approach to resilience makes it possible to offer a framework for dialogue to the various stakeholders, whether these be elected representatives, users, technicians, business representatives, or citizens: "Moreover, these actions have to consider the whole of the exposed and damaged components of a community at the same time (non rivalry) without excluding anyone, no matter what gender or social conditions (non excludability)" (Pesaro, 2018; 3). Through dialogue stakeholders will become aware of each other's imperatives and the importance of collectively developing responses adapted to the realities on the ground. This makes it possible to overcome blockages and create links based on trust between those involved, who can be called upon in the event of a crisis. Collaborative projects should be encouraged, not isolated ones. If resilience is considered common good, then public governance for the development of a regional approach to resilience is fully justified (Pesaro, 2018). This leads to the recognition that there is a market gap: economic players cannot produce the conditions for resilience on their own; the intervention of public authorities (the State or a local authority) is necessary here. The City of Paris' resilience strategy, for example, was jointly constructed through collective intelligence workshops attended by a total of 800 stakeholders. Institutional and economic entities, associations, and citizens have contributed to building an action plan around the "will to build a more inclusive city designed for its inhabitants" (Cerema, 2019a,b). The approach is also based on cooperation with the network of 100 resilient towns and with neighboring municipalities.

The third pillar, entitled "Learning and Innovation," invites stakeholders to engage in a learning process, by asking questions, particularly about past events. But one must be able to break away from this past because the same causes do not necessarily produce the same effects. So we have to ask ourselves the following questions: "Are we able to make different decisions today? Are we safe from making such a mistake again? How can we avoid such damage next time?" Learning and innovation can be brought into play through calls for projects such as the Rhône Plan. Launched in 2017, the Rhône Plan involved local authorities and organizations to raise awareness of the risk of flooding by the use of tactile models, soundwalks, and serious games. The innovations are not only digital, but also organizational, such as the steering committees of the local flood risk management strategy.

The fourth lever, "disaster risk reduction," is based on explaining the "acceptable risk" resulting from a choice between the cost of protection solutions and the acceptability of a level of damage. We know that not all the people involved will give the same weighting to the acceptability criteria. Infrastructure managers, for example, will be attentive to service continuity or resumption, while emergency services will attach importance to the safety of people and security of property. The resilience exercise for this lever will involve making stakeholders reveal their preferences to choose acceptable adaptation and protection measures. Reducing the vulnerability of critical infrastructure and networks is therefore a major lever of resilience during disasters, in particular to help the area recover rapidly. The aim is to reconcile the sometimes conflicting interests between: energy, telecommunications, water, sanitation, and transport network operators; and the relevant local authorities.

The fifth lever, "identity and specific features of the area," is based on the fact that an area that relies on its own resources and protects its common goods finds its resilience strengthened, reduces its dependence on the outside, and guarantees the survival of its inhabitants. To do this, knowledge of the area and how it works is essential to understand how it was built and what structured it. Local culture, the memory of crises, and disruptions are part of its identity: *This means that, starting from a similar territorial capital, the characteristics and the image of the area may deeply differ, producing different exposure and vulnerability profiles* (Pesaro, 2018; 1). By setting the area within the context of its history and projecting one or more possible future scenarios for it, stakeholders can take ownership of it and incorporate it within a larger scale, in keeping with the challenges being considered.

The sixth lever, entitled "Monitoring and anticipation," is based on a forward-looking approach founded on knowledge of the area's past and present to better anticipate any disruptions that may arise and take measures to prevent them. This promotes better organization during major crises and stresses or optimized trajectories in the face of future change. In the event of identified threats, such as volcanoes or landslides, this approach is implemented naturally. It is also useful in the event of poorly defined hazards, or to analyze the interdependencies of risks in situations of complexity and uncertainty. The aim is to think about the scope of possibilities when a planned procedure corresponding to a scenario identified in advance cannot be referred to. In France, the role of the *Observatoire régional des effets du changement climatique* (Regional observatory of the effects of climate change) in Auvergne-Rhône-Alpes is to collect and analyze data and to lead networks of stakeholders so that they incorporate the effects of climate change into their projects.

10.2.2 Methodology for using the resilience framework

For urban development the resilience framework can be used ex ante or ex post. It can be used for the same development by the different parties involved in the urban project, in particular inhabitants and/or users. The use of collective intelligence methods is particularly recommended. Not only the developer but also those involved in the economic setup of the project can be involved: different urban resource management agencies; the project manager: the various companies; the engineering departments of the local authorities in charge of maintaining and managing the public space in which the development is deployed, such as the road department, the water department; users and residents; or members of neighborhood life associations.

10.2.2.1 Ex post: discussion, assigning points, and revealing forgotten issues

In the case of an ex post assessment, the interviewees are asked to list the aspects of the project that do or do not involve resilience for each of the six levers of resilience and each of the three subcomponents.

At the end of this examination, the interviewees give one point for each time at least one subcomponent is fulfilled by the development carried out. A maximum of three points can therefore be assigned to each lever, even if several aspects contribute to achieving this issue. Conversely, if an aspect shows that one of the subcomponents is not taken into account, a minimum of zero point can be assigned. Each lever can therefore receive between zero and three points. The development can therefore receive a minimum rating of 0 and a maximum rating of 18 points. The tool therefore makes it possible to assign a final score to a development. It can also be used to compare it with others. But perhaps the most interesting feature of this tool is its ability to qualitatively assess a development. This is because it shows whether the various initial issues have been reconciled and reveals the complexity of the thinking undertaken.

10.2.2.2 Ex ante: laying the foundations for a resilience strategy

The framework can also be used ex ante to lay the foundations for a resilience strategy. Three stages are necessary for this (Cerema, 2017).

The first stage is a shared diagnosis of the area. The aim is to involve the widest possible collective of stakeholders to build a shared vision of the area. The framework is then used to raise awareness of resilience among stakeholders, share knowledge and a representation of the area; identify local risks, threats, weaknesses, and strengths; and finally select the main issues.

The second stage aims to build a unifying resilience strategy, through a hierarchical action plan that responds to the issues identified in the first stage. Each action or initiative should make use of as many resilience levers as possible, integrate and/or synergize sectoral projects, and choose a few flagship projects.

The third and final step is to implement, monitor, and assess the planned actions in a cross-cutting manner while linking them to the running of sectoral projects. The control of communication about the project, and the monitoring and assessment process for the approach will be based on a choice of indicators.

10.3 The resilience of Place de Francfort in Lyon: an integrated approach, diverse stakeholders, and disaster risk reduction

The resilience framework was used to make an ex post assessment of the level of resilience of a recent development in Lyon: the requalification of

Place de Francfort (Section 10.3.1). We chose to use the resilience framework, given the difficulty of carrying out a CBA of the development. This is because it is extremely difficult to monetize the reduction of urban heat zones and the reduction of runoff. Following an interview with the operation managers of the *Société d'aménagement de la Part Dieu*, responsible for designing and managing the planning of this public space, the framework was used in-house at Cerema by four of the coauthors of this section. The method revealed that this urban development had a high level of resilience (Section 10.3.3).

10.3.1 Place de Francfort: a gateway to Lyon's metropolitan area with high urban challenges

Place de Francfort measures 8500 m^2 (Fig. 10.2). It is the gateway to the Lyon metropolitan area for airport and rail passengers. It is an integral part of the multimodal interchange hub of the Part-Dieu railway station, one of the busiest in France with more than 31 million passengers per year. Over 30,000 pedestrians pass through it daily and it features many hardscape materials that make it a major urban heat zone in the very heart of Lyon. As a result, quality of life and comfort of use are significantly affected during heat waves that tend to occur more and more regularly as a result of climate change. In addition to improving connections between the various means of transport (connections with the bus station, the tramway, the railway station, etc.), improving how it operates and serves all users (passengers in transit, residents, local employees, and tourists), the

Figure 10.2 Photo of Place de Francfort. *Courtsey LDaniere.*

redevelopment of the square must also improve the comfort of pedestrians. The square is divided up into several distinct uses that connect with each other: a drop-off car park for private vehicles, which will eventually be occupied by buildings; a bus station for coaches and a pedestrian square measuring nearly 4300 m^2.

10.3.2 Innovative developments to adapt to climate change

One of the main focuses of the requalification operation was to improve the bioclimatic atmosphere of the square by combining three actions. Action 1 consisted of limiting surface temperatures by using light gray granite paving stones with a higher albedo[1] than black asphalt, the initial surface covering. Action 2 enabled the efficient use of rainwater through an alternative rainwater management system by helping water to infiltrate through a German-made porous joint between the paving stones and preparing a subsoil with a high water absorption capacity, helping to irrigate the trees. Action 3 is based on dense planting of 93 trees separated from each other by 5-m steps and planted in a single container so that their roots can communicate. The tree species were chosen to withstand extreme hot temperatures. Each tree is inserted into a ring that allows people to move around as close as possible to the tree. This ring is designed to expand as the tree grows.

10.3.3 The development Place de Francfort: relatively well-reconciled objectives

We (four of the five people who signed this paper) made an ex post assessment of the layout of Place de Francfort using the Resilience framework. We have listed the factors pertaining to resilience for each of the six levers of resilience and each of the three subcomponents. Our assessment revealed a number of points and a very positive overall evaluation (Fig. 10.3).

Concerning the first lever, "Consider the area in its entirety," we consider that the development carried out takes into account the benefits and cobenefits and that the action is "no regrets" because it benefits the area, apart from the fact that it has also been designed to meet the challenges of climate change. We identify no particular dependency. In our opinion the risks were properly taken into account in the design, since an experimentation table was installed to test several innovative surface covering

[1] Albedo is the power of a surface to reflect sunlight.

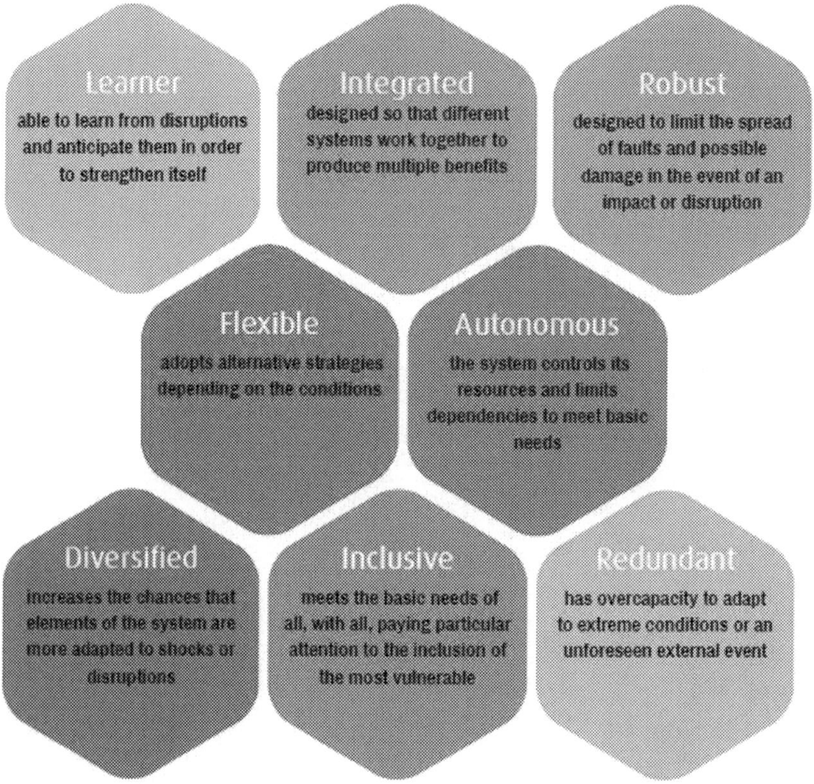

Figure 10.3 Description of the six levers of resilience.

and planting solutions. The materials were subjected to hot temperatures and various uses, including the movement over them of people with reduced mobility. Soil studies (permeability and groundwater level) highlighted the feasibility of rainwater infiltration and contributed to the design of the structures. And urban security issues, including the risk of terrorism, led to a choice of tree species that do not hide cameras. Our assessment is therefore 3/3 for this lever.

For the second lever, "stakeholder diversity and cooperation," we give a positive assessment of the work that the developer has done upstream with the departments that will carry out maintenance of water, roads, and green space, and that done with associations of users with reduced mobility. For tramway crossings the same surface covering as in Place de Francfort was used to guide users to the bus station. For these reasons, we gave a score of 3/3 for this lever.

Concerning the third lever, "encourage learning and innovation," we consider that the urban planner could have done more by displaying informative posters on the site. We know that when knowledge is nonexistent or incomplete, it is necessary to experiment, and to adopt flexible means of action, which can be adapted over the long term according to unforeseen events. The urban planner was able to adapt and manage uncertainty by testing the albedo of the paving stones, modeling the urban heat zone, or assessing the comfort of wheelchairs running on the materials used. It strongly encouraged its suppliers to innovate. Our assessment is 2/3.

For the fourth lever, we believe that disaster risk and exposure have been reduced. This is because, in the event of runoff, soil studies have demonstrated the feasibility of rainwater infiltration and have helped to design the structures. In addition, in the event of heavy rains, the square can still be used because of a slope that discharges water into the drains. However, we highlighted certain concerns, such as the possible case of joint clogging because of dirt on the ground and possible trees falling in the event of a storm. How responsive would the services be? Nevertheless, technical robustness is quite good because the solutions are not based on a complex technological system. Redundancy means the possibilities of rerouting pedestrian in the event of the square being closed and the fact that there are two different entrances to the station. Our rating is therefore 2.5/3.

For the second fifth lever, "assert specific features of the area," we regret the fact that no aspect affirms the local features of the place, especially since the paving stones come from Germany. We know that resilience also depends on the preservation of local commons. Here the groundwater table is replenished by rainwater infiltration and the ground is designed to retain large particles of pollution which is very positive. Our rating is therefore 2/3.

Concerning the sixth and final lever, "promote monitoring and anticipation," our assessment is very positive and we give a rating of 3/3. This is because many diagnoses of the area have been carried out to design actions suited to the environment where they will be carried out: urban heat zones and the choice of trees that can withstand heat. Runoff studies to select joint permeability with an estimate of the level of rainfall intensity with which the site is potentially confronted have been carried out. In addition, consideration has been given to crisis and postcrisis preparedness. If there is a shock on Place de Francfort, pedestrians could be rerouted.

The ex post assessment of the requalification of place de Francfort using the Resilience framework gives a score of 15.5/18.

10.4 Conclusion

Our primary objective was to produce a tool to help better understand the levers that promote resilience dynamics.

We achieved this by means of a framework that was administered in-house at Cerema and with the help of the urban planners of the Part-Dieu development company in Lyon. On the basis of this first assessment work, we are able to present the factors that make up the resilience of the Place de Francfort.

This assessment is only one illustration of what can be done with the resilience framework. The tool was not used according to the methods of collective intelligence, which would have allowed the various stakeholders to compare their points of view. All these steps should be carried out to complete the assessment of Place de Francfort. A "world café" type workshop could be useful. During this workshop, each of the six levers would be presented on six different tables. The various stakeholders would share their assessment of the development with respect to each of the levers by moving from one table to another. A group leader could report on the proposals made by the previous group, thereby allowing those present to draw inspiration, to react, and go into the assessment and proposals for improvement in more depth.

The results obtained using the resilience framework confirm our approach. We believe that conducting an assessment requires indicators to be produced to objectify resilience. Insofar as there is no standard or reference value, apart from regulation, on the levers identified by this tool, the assessment conducted here can produce only a discourse, or qualitative assessment. Nevertheless, the resilience framework remains operationally relevant. The tool has a hermeneutic function. It is there to question upstream or downstream. It also makes it possible to share knowledge and clarify political decisions as to the degree of risk, stress, or dependency accepted for the development of the operation in question. In this way, it makes it possible to objectify decisions.

References

Bruijne, M., Boin, A., & van Eeten, M. (2010). Resilience: Exploring the concept and its meanings. In L. K. Comfort, A. Boin, & C. Demchak (Eds.), *The rise of resilience, designing resilience. Preparing for extreme events* (pp. 13–32). Pittsburgh, PA, USA: University of Pittsburgh Press.

Cerema. (2016). *Urban resilience and crisis management.* Lyon, France: Cerema.

Cerema. (2017). Résilience urbaine et sécurité des territoires. Crises redoutées, résiliences escomptées et étapes à franchir [Urban resilience and regional security. Crises feared, expected resilience and steps to be taken]. In: *Proceedings of the 2017 Cerema workshop.* Lyon, France: Cerema, (in French).

Cerema. (2019a). *Sécurité globale et résilience des territoires Ingénierie, effets de levier et stratégies à promouvoir [Global security and territorial resilience. Engineering, leverage effects and strategies to be promoted].* Lyon, France: Cerema.

Cerema. (2019b). *Vers des territoires résilients – six leviers opérationnels pour anticiper, réagir, s'adapter, Fiche méthodologie n°01 [Towards resilient areas – six operational levers to anticipate, react, adapt, Methodology factsheet no. 01].* Lyon, France: Cerema. (in French).

Cutter, S. L., Barnes, M., Berry, M., Burton, C., & Evans, E. (2008). A place-based model for understanding community resilience to natural disaster. *Global Environmental Change, 18*(4), 598–606.

Hatchuel, A., Labourdette, A., Leduc, F., Pillaud-Vivien, L., & Renaudin, M. (2017). *La Résilience à Paris: états des lieux et préconisations multi-bénéfices pour l'espace public [Resilience in Paris: Inventories and multi-benefit recommendations for public space].* hal-01824659, (in French).

Hernandez, J. (2009). The long way home: une catastrophe qui se prolonge à La Nouvelle-Orléans, trois ans après le passage de l'ouragan Katrina [The long way home: A continuing disaster in New Orleans three years after hurricane Katrina]. *Espace Géographique, Tome 38*(n°2), 124–138.

Holling, S. C. (2001). Understanding the complexity of economic, ecological, and social systems. *Ecosystems, 4*(5), 390–405.

Pesaro, G. (2018). The contribution of economic thinking to innovate disaster risk reduction policies and action. In Colucci Galderisi (Ed.), *Smart resilient and transition cities, Emerging approaches and tools for a climate-sensitive urban development* (pp. 249–256). Amsterdam: Elsevier.

Pigeon, P., & Rebotier, J. (2016). *Disaster prevention policies. A challenging and critical outlook.* London: ISTE/Elsevier.

Sen, A. K. (1985). Well-being, agency and freedom - the dewey lectures 1984. *Journal of Philosophy, 82*, 169–221, In press.

Timmerman, P. (1981). *Vulnerability: Resilience, and the collapse of society: a review of models and possible climatic applications.* Toronto: University of Toronto, Institute of Environment Studies.

CHAPTER 11

When the community takes action: building resilience through the Transition Movement in Spain. A critical assessment

Cristian Moyano-Fernández[1,2], Salvador Garrido-Soler[1], Juan Del Río[1,3], Guillermo Rodriguez-Navas[1] and Lisa Cassanet[1]

[1]Red de Transición, Spain
[2]Universitat Autònoma de Barcelona
[3]ECOLISE, Spain

11.1 Introduction

If we were to briefly explain the evolution of the Transition Movement (TM) in Spain to anyone, we would invite our partner to read these two quotes and look for the difference between the one and another:

"We give the impression that we do not want to know that our prosperous industry and our comfortable life depend on reservoirs of fossil fuels that before a hundred years will have run out. The problem, in the near future, will not lie in making new prospections and opening new trial pits. But if the case arises, man will be able to boast of a new achievement in this era of breaking records: having drunk in a century the wealth that took 600 million years to form" (Delibes, 1979: 98).

"We are witnessing the chronicle of a death foretold: the vanishing of nature. The media continue to make up the seriousness of the environmental deterioration, always trying to accompany the news of unusual climatic phenomena with some reference to something similar that happened in the remote past. This makes it difficult to perceive that we are already facing a dangerous global climate change generated by an irresponsible techno-economic model" (Puleo, 2017: 211).

Almost 40 years separate both quotes and their content is unfortunately similar and current, but the biographies of their authors are

[ECO]systems of Resilience Practices
DOI: https://doi.org/10.1016/B978-0-12-819198-9.00002-8

completely different from each other. The first was written by a conservative, journalist, writer, and conservationist man. The second belongs to a progressive woman, researcher and lecturer, and ecofeminist. Their comparison allows us to perceive the evolutionary change that Spanish society has suffered during the last decades and notice the role of TM in this evolution. On the following pages, we are going to explain this process.

11.2 Preliminaries

11.2.1 Socio-ecological movements background in Spain

The Spanish environmental movement, in its beginnings during the late 1960s and in the early 1970s, was formed by pioneering groups (SEO, ADENA, AEORMA[1]) dedicated to protect the natural environment of animals, landscapes, and land use planning. The concern of these movements about the social and political spheres increased later on, criticizing some practices and trends rooted in an increasingly neoliberal society (neighborhood mobilization, antinuclear protests, against pollution or concrete-intensive construction policy). Marxist culture, the struggle for democracy, disaffection toward traditional parties, nationalism, and regionalism were defining references of Spanish ecologism, yet with their own styles (Fernández, 1999: 46–47).

Furthermore, due to the political ecology, internationally reinforced by Murray Bookchin, Alain Lipietz, or André Gorz after "The Limits to Growth" published in 1972, and in the Spanish context thanks to the spark ignited by Petra Kelly in 1983 through the so-called Tenerife Manifiesto, the political thinking in Spain included ecological considerations in its programs too. Although the green ideology was already rooted in the Spanish society at the time, it had not jumped to the political activism yet, using the channels enabled by the political system, via a green political party.

For some authors, some important differences make the Hispanic green ideology and the context of the new social movements a particular case in Europe (Piulats, 1992: 139–155). To understand the ideology of the Spanish environmental movement, one should look back at what

[1] Spanish acronyms, respectively, of the Sociedad Española de Ornitología, Asociación para la Defensa de la Naturaleza and Asociación Española para la Ordenación del Medio Ambiente. Subsequently, SEO became the Spanish subsidiary of Birdlife and ADENA of the NGO WWF.

Iberian anarchism and the phenomenon of libertarian communes meant (Puente, 2003). These movements have followed the path of eco-socialism and political ecology for the last decades, rather than either a spiritualism based on the deep ecology philosophy of Naess and Sessions or an ecofeminist movement.

Nevertheless, numerous models of intentional community exist or have existed in Spain. In the face of the consumerist individual, politically submissive and responsible only for himself, there are people who consider that the community is the best option for a responsible and realistic freedom (Escorihuela, 2012). It is likely that the hippie philosophy that came from America and landed in Spain during the 1970s, with its first representatives in Ibiza (e.g., the so-called Peluts), and their arrival to Granada to "free the land",[2] helped to crystallize these new forms of social organization in communes and personal attitude based on nonviolence.

The first ecovillages arrived in Spain in the early 1980s with different motivations and origins (e.g., Lakabe in Navarra, Los Portales in Seville). Further ecovillage models started from meditative life and the search for spirituality as main axes. Although the origin and approach of each ecovillage are different, almost all people embarked on such projects have in common their idealism in response to a nonconformity with conventional society, enormous creativity, love for nature, and inner search. Despite their wide diversity, every community respects the common aim of creatively transforming one's life and the environment toward new self-managed coexistence models, fostering regeneration.

Moreover, the beginning of the agroecological movements in the 1990s and its development during the entry of the 21st century were incorporated in many of the communes, ecovillages, and alternative societies that were looking for a new design and more sustainable management of the crops. The arrival of permaculture (Ross, 2005) as a holistic method to work with nature contributed to experience new collaborative ways of treating the fields, designing the architecture of the houses, and interacting with others and with the environment.[3]

All this background built a complementary horizon to facilitate the TM flourishing.

[2] https://www.elmundo.es/sociedad/2016/05/25/57449488e2704e80758b4650.html.

[3] Consulted in: http://www.tierramor.org/permacultura/PermahistoriaInternacional.htm#des. Permaculture movement arrival can be exemplified by the Instituto Permacultura Montsant in Tarragona: http://www.permacultura-montsant.org/.

11.2.2 The birth of Transition Movement in Spain

TM was born in 2006 in Totnes (United Kingdom) and is currently present in more than 50 countries. As a flexible model, it is being adapted and used by communities with different cultures and realities. They are organized to create resilient societies, capable of adapting and absorbing ecosocial shocks and moving toward a more local and self-sufficient future, less dependent on oil and other nonrenewable resources.

In Spain the first TM groups were formed in 2008, with Barcelona, Zarzalejo (Madrid), Coín (Málaga), Jerez, and Mancor de la Vall (Mallorca), as some of the pioneers (Del Río, 2019: 86). 2008 was precisely a year marked by the financial crisis triggered by the collapse of the real-estate bubble in the United States. The beginning of the global economic crisis meant for Spain the explosion of other problems: the end of the housing bubble, the 2010 banking crisis, and the unemployment increase, which led to the social movements emergence aimed at changing the economic and productive model as well as questioning the political system, by demanding a democratic renewal (Mella, 2013).

Neoliberal policies and the financed economy propelled by the Welfare State in the 1980s silenced the first ecological criticisms born in Spain, which denounced the material limits with which capitalism would collide. Economic growth did not stop, and the quality of life of the Spanish population increased. So why be worried? The housing boom in Spain had similar characteristics as in the United States, driven by liberal bank and mortgage institutions lending and high mortgage products, followed by securitization of the markets (Casais, 2011: 37–62). The "cajas," smaller or regional banks that lent money to families in Spain, were lacking liquidity because of the indebted households and the construction industry which owed billions to the Spanish banking system (Fogg & Harrington, 2011: 51–65). But this speculative bubble formed between 1994 and 2007 burst causing an increase in unemployment, a reduction of consumption, a weakening of the construction sector, the closure of up to half of housing agencies, and an instability of financial institutions, leading to bank mergers to ensure the survival of some of them.[4] Thereby, the banks initiated a strategy of forced evictions that left hundreds of thousands of Spanish people to lose their homes. This critical situation gave rise to numerous protests, organizations, and mobilizations that

[4] It can be consultated in https://elpais.com/diario/2008/01/17/economia/1200524403_850215.html and in https://elpais.com/diario/2009/02/22/economia/1235257202_850215.html.

would demand an improvement in living conditions, violated after the crises started in 2008. Probably, the most important social movement was the so-called 15-M Movement in 2011,[5] which arose due to the precariousness and economic conditions of the middle and lower classes, two consequences of the financial crisis.

There is a direct relationship between the economic crisis context and the increase in the number of initiatives (Chelleri & Del Río, 2017). This historical scenario could have been an incentive for the rise of transition initiatives (TIs) toward a more sustainable, communitarian, and self-sufficient way of life. The discontent of some inhabitants of the Spanish population regarding the growing economic precariousness, the social conflicts resulting from a competitive system full of inequalities, and the environmental impact of the neoliberal lifestyle motivated the creation of new ways of living in community. As indicated previously, the origin of the first environmental associations was hippie communes and ecovillages, as well as the adoption of ecological agriculture to guarantee food sovereignty and holistic principles of permaculture. Some like-minded people from these movements embraced the transition philosophy and started to set up their initiatives. Some communities and projects already contained many of the principles of the transition implicitly even if they were not named TIs. But that was irrelevant, because with their actions they were already moving toward a world of transition, resilience, and regeneration. So, the aforementioned background of socio-ecological movements sprouting in Spain, a few decades before the arrival of TM, already suggests that TM is not a closed path but open and diverse, with the presence of bottom-up processes. It is the people who organize collectively to design a different future through local initiatives.

11.3 Current status of transition initiatives in Spain

11.3.1 General retrospective since 2014

After the first study dedicated to clarifying the scope of the TM in Spain (Chelleri & Del Río, 2017), important political, economic, social, and

[5] This is one of the most significant social movements to emerge in Spain since the restoration of democracy, characterized by the acquisition of reticular, horizontal, or participatory forms of organization, the result of a historical process that was independent but simultaneous with the development of the Internet (Candón & Redondo, 2013: 117). However, the "family tree" that led to its emergence includes, unlike the TM, the countercultural movement of 1968, the alter globalization movement and the movement of the indignant as the most influential references (Ruiz-Aja et al., 2013: 49–61).

cultural changes took place throughout the country. The main one is the timid economic recovery after the dramatic program of budgetary adjustments faced during the Great Recession (2008–14). Its devastating effects in the southern countries of the European Union were not accompanied by a social and cultural recovery. However, adjustment policies (or austericide, according to the expression coined by critical economists) provoked an unprecedented increase in social inequality and triggered a period of political instability and territorial crisis that still persists.

With these antecedents, it was necessary to know the situation of the TM in the Spanish State. A university internship carried out by the coauthor Lisa Cassanet, applying the methodology approved by the Spanish Transition Hub,[6] enabled a new investigation into the TM situation during the period 2014–17. The objectives of this research were (1) increasing the visibility of active local initiatives, (2) improving their connection and networking with each other, with the Spanish Transition Hub and local administrations, and (3) diagnosing their internal functioning to suggest improvements (Red de Transición España, 2020). These objectives were consistent with the needs identified in the previous 2014 report and entailed an adaptation of the methodology to cover a broader spectrum of the structure and functioning of local TIs.

The preliminary conclusions of the second study on the situation of the TM in Spain were presented at a Strategic Meeting of the Spanish Transition Network, held in January 2018 in the town of Valls, Tarragona (Red de Transición España, 2020). The study included four differentiated parts, allowing accurate diagnosis of the real state of the TM in Spain: general situation of local TIs (part 1); objectives, main themes, and projects (part 2); challenges, barriers, and needs (part 3); and TIs as part of a general change movement (part 4). An overall view of the TM evolution can be extracted from a comparison between the two studies.

11.3.2 Nature and evolution of the initiatives

The definition of TIs is one of the differentiating factors between the two studies. In 2014 the difficulty of specifying the true scope of the

[6] Transition Hub Groups and organizations form at various levels of scale to catalyze and support TM. These Hubs are as diverse as the communities they arise in. A National Hub can be particularly useful when other parts of social society want to engage with TM at that level of scale. They are a distinctive layer in the TM, connected by regular international gatherings, communications, international working groups, and a very strong shared sense of purpose. The Hubs Circle and Working Groups make decisions by consent within circles, using sociocratic way (Transition Network, 2020).

movement was noticed. For this reason a distinction was made between TIs as such versus other initiatives with principles or activities (degrowth, permaculture, ecovillages, integral cooperativism, new economies, complementary currencies, or self-managed social centers, among others). The list of both initiatives served to warn about the local TIs proliferation during this period and to verify the heterogeneity of these movements (Chelleri & Del Río, 2017: 33).

In 2018 the scope of the study was delimited to the local initiatives self-denominated as such; in other words, those that were recognized as part of the TM carried out activities linked to the transition or that included the appellative "in Transition" in their denomination. All local initiatives that agreed to participate in the study were asked to provide their own definition of what they understood to be a local TI. The objective of this question was twofold: to establish a precise conceptual framework on the TM nature and to know the self-perception of each local initiative as a group.

The results of this initial question were very varied and reflected different conceptions about the TM. The construction of a common definition based on terminological analysis could be approximated with this statement: a TI is a human group that responds to the social and ecological crises by promoting community resilience and self-sufficiency, to create sustainable living alternatives through social awareness. A local TI rejects the conventional system of production and consumption and seeks to recover exchange, cooperation, and the common good. It is characterized by being degrowth, ecologist, feminist, collaborative, supportive, and democratic.

In the light of this meaning, it is interesting to note the evolution of movement through some relevant figures. An amount of 76 initiatives arose between 2008 and 2017, as calculated in the latest 2017 report. The annual rate of local initiatives creation (Fig. 11.1) was used as a reference indicator to study the historical evolution of the movement: the greatest increase in 2012 and 2015 and the declining trend from 2015 onwards. The holding of meetings or other connecting activities (e.g., State or Regional Transition Meetings), as well as territorial campaigns that promote or adapt the principles of transition to a more concrete context (e.g., Resilient Andalusia Campaign in 2016) are determining factors in encouraging groups to create a new TI. On the opposite side, periods without projects or activities to connect or support initiatives suggest a lack in the emergence of new local groups.

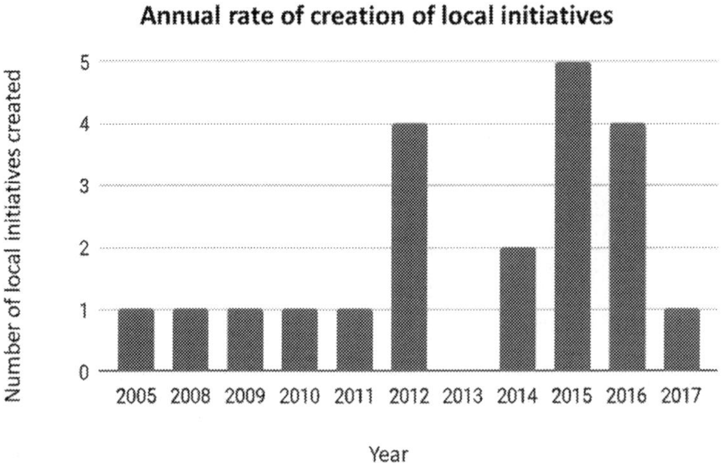

Figure 11.1 Annual rate of creation of local initiatives. *From: Red de Transición España. (2020)*. El movimiento de transición en el Estado español. Actualización del mapa de iniciativas locales de transición (2014–2017*)*.

However, this global count of 76 initiatives is not representative of the movement. It seems more appropriate to stick to the actual level of participation of local initiatives in each of the periods studied. In 2014 there were 18 TIs responding to the questions posed in the interviews, while in 2017 there were 26 TIs. This low level of participation corroborates the hypothesis put forward in the 2014 study: the participating initiatives are the most active in the movement, although there are others whose level of internal organization did not allow them to answer or participate in the study (Chelleri & Del Río, 2017: 36). The updated list on the evolution of the movement in Spain included the following categories (Fig. 11.2): active initiatives, inactive initiatives, sleeping initiatives (currently inactive but capable of reactivation), and unlocatable initiatives (inactive, disappeared, or which have declined their participation). From this list, it is possible to verify the state of the movement, totaling around 20 initiatives in operation.

The temporary evolution of the movement is also complemented by a territorial and thematic analysis. With regard to the territorial distribution of local initiatives, there is a progressive concentration of TIs located in areas close to large cities (Barcelona, Granada, Madrid, or Tarragona), probably due to institutional or financial support factors. And regarding the main themes of the activities developed by TIs, in 2014 it was noted

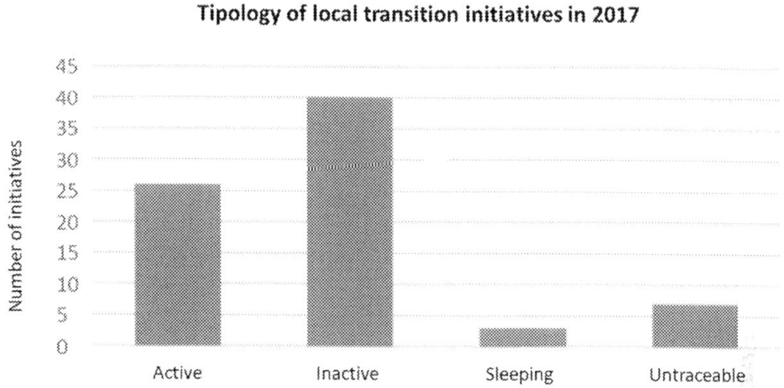

Figure 11.2 Situation of TI in 2017. *TI*, Transition initiative. *Red de Transición España. (2020)*. El movimiento de transición en el Estado español. Actualización del mapa de iniciativas locales de transición (2014–2017*)*.

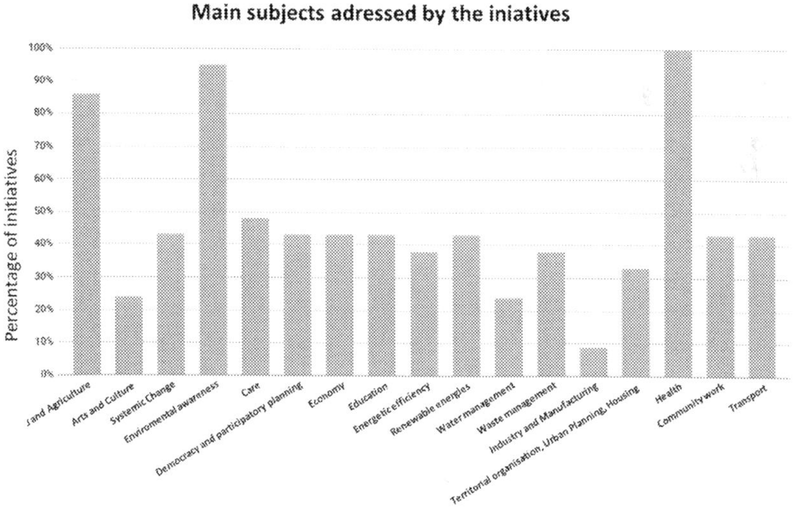

Figure 11.3 Projects and activities of Spanish TIs. *TIs*, Transition initiatives. *Red de Transición España. (2020)*. El movimiento de transición en el Estado español. Actualización del mapa de iniciativas locales de transición (2014–2017*)*.

that education and food dominated 40% of the experiences collected (Chelleri & Del Río, 2017). Later, in 2017, it was decided to present a complete list with 17 topics so that the groups could determine the profile of their initiative based on their activities and preferences (Fig. 11.3). The results determined three central axes (health and quality of life, environmental awareness, food, and agriculture), although these groups recognized the need for inclusive and open projects to raise awareness throughout society.

11.4 Challenges and opportunities of Transition Movement

The internal dynamics of TIs and their groups in charge is the most sensitive area of both studies. Exposing internal problems and organizational difficulties is always complicated, because it shows the limitations as groups that hinder their sustainability. However, both reports remark the transparency and honesty of the participating groups and similar patterns are observed in both. The 2014 study inquired about the main barriers and difficulties faced by the initiatives. The results revealed three problematic areas (Chelleri & Del Río, 2017: 42—43): (1) lack of funding and financial support, (2) lack of participation and internal coordination difficulties, and (3) lack of communication between different groups and connection with other initiatives.

In response to these difficulties, the 2017 study included these concerns as the main objectives. Many questions addressed in greater detail the needs, difficulties, and challenges for TIs. Additional interviews carried out on the basis of the questionnaire results made it possible to delve deeper into the most problematic aspects (Fig. 11.4): the main challenge faced by the TIs is not the lack of funding, but of participation.

There is a permanent lack of people and time to involve more actively in the many projects at the same time. This may be explained because a TI provides a space in which volunteers devote large amounts of energy and time to transition-related activities or to the coordination needed to maintain the activities and the initiative structure. The amount of time invested in organization and coordination barely leaves time for communication tasks that guarantee the incorporation of new people.[7] Relief and new incorporations are fundamental to guarantee the stability of the projects. In addition, the need to convey illusion to achieve this objective is pointed out, also the suitability of carrying out small successful projects to encourage people and avoid their frustration. In a movement that is

[7] The problem of communication to involve more people is a recurrent problem within current social movements, called also "communicative endogamy" (Jiménez-Gómez & Olcina-Alvarado, 2017: 13). To overcome this problem, these authors recommend introducing communication policy in all activities of a social organization.

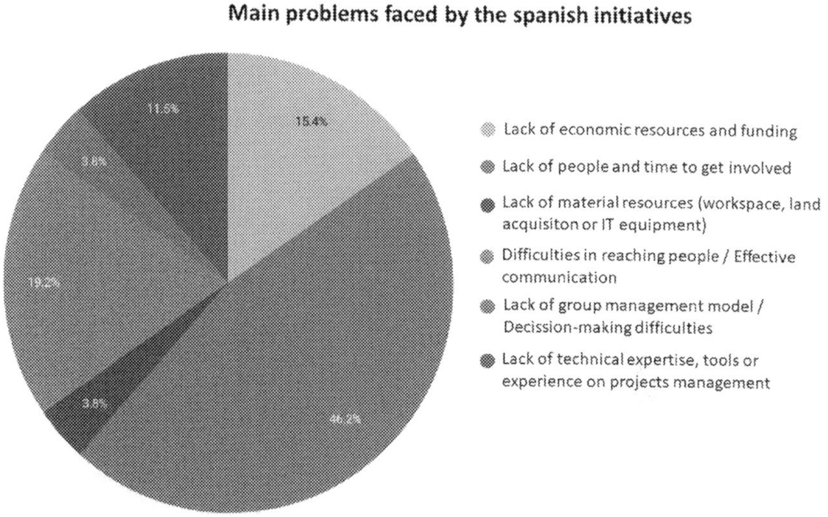

Figure 11.4 Main problems faced by the Spanish initiatives. *Red de Transición España. (2020)*. El movimiento de transición en el Estado español. Actualización del mapa de iniciativas locales de transición (2014–2017*)*.

especially concerned with the care of people and their emotional stability, these are particularly significant conclusions.

In these conditions, it is very difficult to consolidate an initiative nor the stability of projects in the long run. This conclusion was already pointed out in the 2014 study, but on this occasion an important difference is noticed. If then the revolutionary power of self-organization was pointed out, which makes it possible to break with previous dependencies and the institutional rules they entail (Chelleri & Del Río, 2017: 42), now the initiatives would recognize the importance of collaboration with the public administration. Fluid relationships with local institutions and governments favor the visibility of initiatives, facilitate the diffusion of activities, allow access to public financing and gain the trust of new people and entities.[8] There is no consensus on this issue, since some of the participating initiatives recognize that these relationships are not peaceful,

[8] Here is precisely relevant the Municipalities in Transition project and framework (Municipalities in Transition, 2018). It is a wide initiative allied with the Spanish Transition Hub which aims to develop and test a structured way for municipalities and local governments to connect with their communities and respond to the great challenges of this historical period adopting a systemic view and methodologies inspired by the Transition Towns movement approach and learnings.

involve many bureaucratic obstacles, or openly distrust the relationship with the public administration. As a consequence, it may result on a lack of public funding or public institutional support (Chelleri & Del Río, 2017: 43, Figure 2.2.9).

The connection of the initiatives with the rest of the TM or with other related movements is extremely complicated. The final diagnosis of the report reveals the existence of a score of initiatives with an average of 10–20 people actively working on awareness-raising, food, and health projects. The life cycle of each initiative spans 3–4 years. However, the success stories point out to a specific profile: human groups with a long trajectory and previous experience, a dynamic group whose participants are renewed, rather close relationship with the public administration and sufficient funding that comes mainly from the group itself with external support, including access to land to support long-term agroecological projects.

Besides that, the 2017 study more deeply investigates the connections between Spanish TIs and other related movements. These related groups are those involved in the reconstruction of community resilience and in the construction of a more sustainable, fair, and happy society. Although they do not define themselves as "transitional," this relationship is considered essential and very necessary. The most recent initiatives recognize difficulties in being visible and coordinated, whereas more consolidated ones share resources and even act as umbrellas of other related groups (Fig. 11.5).

Regarding the relationship with the Spanish Transition Hub (Red de Transición), it is detected that the collaboration between both is inexistent or very low for 62% of the initiatives consulted (Fig. 11.6). The lack of time and people involved causes the group's energies to be focused on their own activities management, which limits the possibilities of relating to other initiatives or to Spanish Transition Hub.

Here it is important to critically note some concerning aspects about the inner transition by Spanish TIs. As the data collected by both studies show, on the one hand, there is a scarcity of time among collaborators of TIs, and on the other hand, not every initiative includes enough time for "metareflection," self-organization, or other issues beyond outer particular actions (Fig. 11.7).

Those groups who had a more precise and well-defined internal organization were considered effective in their actions and activities. They explained that less time is wasted when defining roles for everyone and

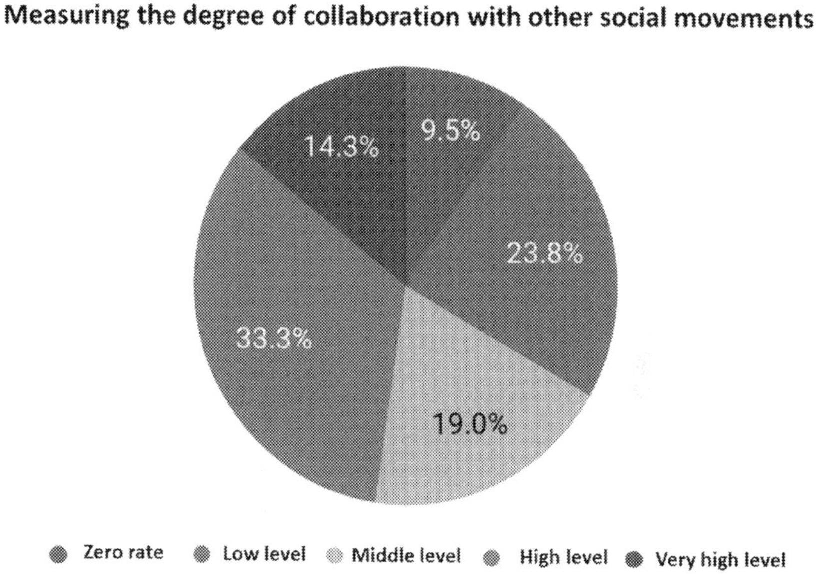

Figure 11.5 Measuring the degree of collaboration with other allied movements. *Red de Transición España. (2020*). El movimiento de transición en el Estado español. Actualización del mapa de iniciativas locales de transición (2014–2017*).*

when it was known precisely which goals had to be followed. This reinforces the perception about how important it is to dedicate time rethinking methodological issues or clarifying the purposes.

Nevertheless, inner transition is broader than methodological issues or just focusing on making the initiative more efficient. It is also essential to take care of the group members. The last studies about Spanish TIs did not inquire nor analyze deeply how this nature of transition is embraced by the groups. Why should it be relevant to embrace a wide inner transition within the daily agenda of initiatives? Because it is necessary to ensure a balance of activity and rest, being aware of the early symptoms of burnout and don't shy away from seeking support sooner rather than later (Hopkins, 2011). Inner transition is helpful for those situations in which our external actions, the effect of our "hands" (following the central principles of the TM philosophy), do not reach what is expected by the "head" or felt by the "heart." If we assume that one of the main difficulties for the members of the initiatives is the lack of time to get more involved, then the internal work becomes crucial. If these circumstances were used to foster inner transition, working resilience, safe shared spaces for freedom of expression,

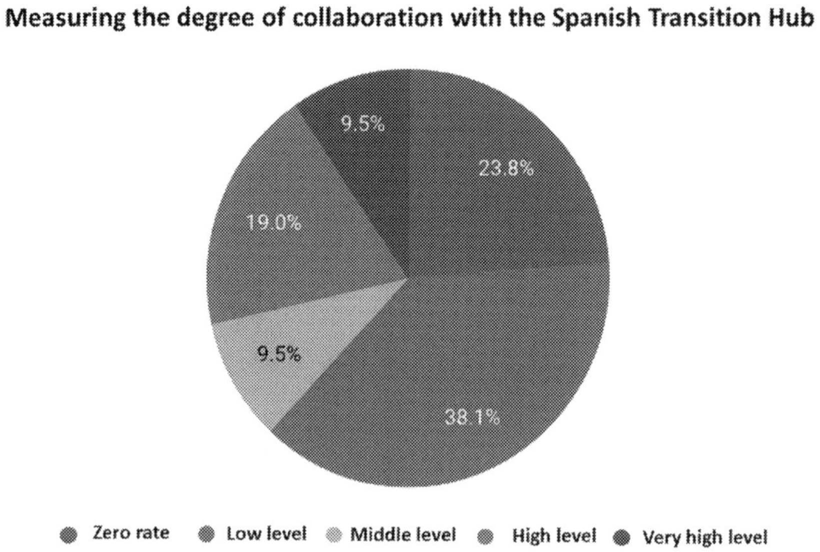

Figure 11.6 Degree of collaboration between local TIs consulted in 2017 with the Spanish Transition Hub. *TIs*, Transition initiatives. *Red de Transición España. (2020)*. El movimiento de transición en el Estado español. Actualización del mapa de iniciativas locales de transición (2014–2017*)*.

support, understanding, and other values associated with the heart, then perhaps the lack of time rather than a barrier or a limitation could be emerged as a reassessment opportunity. In addition, one of the advantages of this shift toward greater consideration of the inner transition may be that it prevents burnout problems.

However, when it was asked to TIs about the level of conflict in the groups, it was very interesting to note that no one defined it as high (Red de Transición España, 2020). In fact, celebrations, parties, and moments for dancing, singing, or food-sharing are considered crucial and they are carried out by many TIs. Daring to dream, dedicating time for building positive vision of future, and cocreating socio-ecological and resilient alternatives are part of TM philosophy too. But whether this kind of celebration is sufficient to fulfill inner transition process is still an open question. The lack of specific data regarding such issues does not allow us to solve our doubts about this. Inner transition also includes reflecting on death, sadness, loss, relinquishment, and other emotions that are seldom expressed in most celebrations. When something happens to us or to the planet that outrages us,

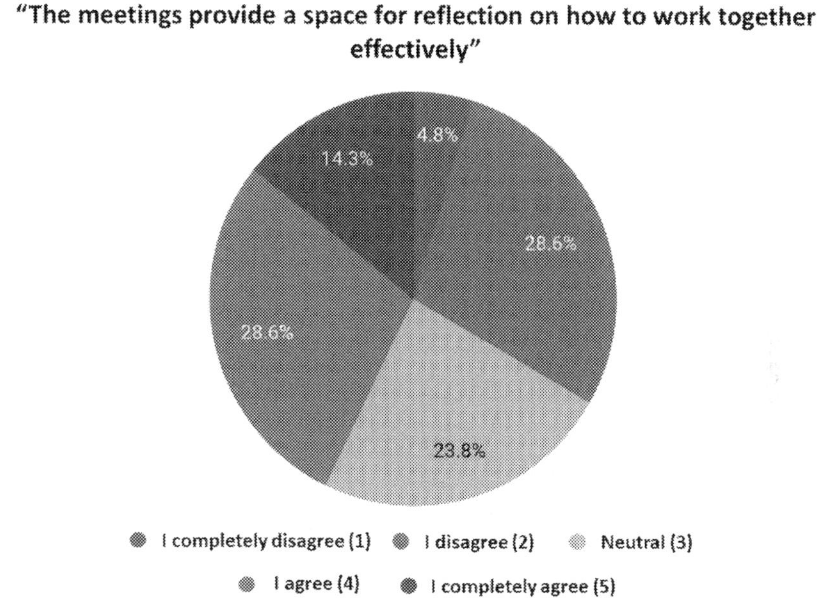

Figure 11.7 The meetings provide a space for reflection on how to work together effectively. *Red de Transición España. (2020)*. El movimiento de transición en el Estado español. Actualización del mapa de iniciativas locales de transición (2014–2017*)*.

how do we channel these feelings (Macy & Johnstone, 2012: 95–112)? Letting space for silence or allowing "retreats" during work or volunteering in an initiative are processes that could also be helpful.

11.5 New trends

Globalization is not only an economic, political, economic, or technological process, but also a cultural one (Marina & Rambaud, 2018: 495). But a wave of deglobalization and postdemocratic authoritarianism is nevertheless sweeping the world in recent years, as repeatedly emphasizes some authors about the future of Western societies.[9]

The emancipatory torch relay in the face of this unclear and turbulent present has been lifted by an unexpected lead character: the world's youth. A new cycle on global social movement has raised two main issues: climate emergency and gender equality.

[9] Source: https://www.publico.es/public/josep-ramoneda-anem-cap-autoritarisme-postdemocratic.html.

The global environmental movement called Extinction Rebellion (XR) was established in the United Kingdom in May 2018 and expanded to Spain between the end of 2018 and the beginning of 2019. During this time, XR has already performed some symbolic actions in Spain: it premiered in Madrid on April 15, when dozens of activists blocked the entrance to Repsol headquarters wearing a banner with the message "There is no Planet B" and denouncing the polluting activity of the company and the criminal negligence of the governments that have been allowing it. Their protests through the massive "die-in" performance are also a common practice of this movement.

Inspired by grassroots movements such as Occupy, Satyagraha, the suffragettes, Gene Sharp, and the civil rights movement, XR wants to rally support worldwide around a common sense of urgency to tackle climate breakdown and the sixth mass extinction. It states the aim of using nonviolent civil disobedience[10] to compel government action to avoid tipping points in the climate system, biodiversity loss, and the risk of social and ecological collapse through protest for global claims, unlike TM which is focused on constructive vision and action, not so much in demanding critically a change.

In a line of criticism similar to XR, the Fridays for Future (FFF) movement has grown greatly in the last year, thanks to the 16-year-old Swedish student Greta Thunberg striking every Friday since the summer of 2018 in front of the Stockholm Parliament. This international movement that brings together the complaints of young people who follow a student's strike to denounce that their future is at stake arrived at the beginning of 2019 in Spain starting with Catalonia.[11] The core of their protests is also to demand that the politicians act for climate change. At the moment the FFF mobilizations are having a positive reception by the mainstream media, but also among governments. Actually, it is estimated that during 2019, there have been 277 strikes in Spain.[12]

The rising of XR or FFF has meant a rupture of the traditional framework with respect to previous social movements. What though makes movements like these so successful and well received in Spain? At the end of 2019, in just 1 year since the appearance of both movements in the

[10] Source: https://xrebellion.org/s/A-Declaration-of-International-Non-Violent-Rebellion-Against-the-Worlds-Governments-for-Criminal-Ina.pdf.

[11] http://juventudxclima.es/2019/02/25/el-fridaysforfuture-llega-a-barcelona/.

[12] Source: https://www.fridaysforfuture.org/statistics/map?c = Spain&d = All&o = all.

Spanish territory, there were around 10,000 followers in the social networks of the first group and about 5000 in the second group. If TM can learn something from both these movements, it would be the efforts they put into organizing collectively under a common goal. The great organizational and management capacity of XR and FFF and its fast impact on the Spanish territory, with a visible influence among young women, exemplify a successful way to involve the community and forge countless networks along the country.

Although all these movements look for similar results, they contain specific features that do not make them the same. One of the biggest differences among these movements is that XR and FFF are focused on global changes, while TM is more addressed to local changes. The famous slogan "Think global, act local" is generally embraced by all the social and environmental movements but maybe TM shifts more weight into the local part, and XR and FFF in the global part. Both kinds of groups start from collective efforts to achieve their goals, but the former focuses on a local scenario with particular actors involved and the latter focuses on a wider scenario with institutional and political actors involved. Also a particular feature of TM is the recovery of ancient knowledge, often considered a suitable know-how to boost local resilience. And although the youth of XR and FFF members justifies its powerful impulse, energy, and sense of emergency (Barrero, 2019), the scope of their claims, and even the accuracy of the scientific data provided in its activities are doubtful.[13]

Other differences are the strategies and attitudes of these movements: the transition projects are positive, constructive, cocreative and with spaces to reflect and celebrate, unlike the other movements based on protests, complaints, civil disobedience, and critiques. This is because their focal points are located at different time scales: the dangerous and deterministic situation of the present, versus the open situation of a future full of hope.

Precisely, the aspect of hope could be a tipping point that can navigate toward the transition philosophy or toward what is becoming increasingly known as deep adaptation (DA). DA proposes to start from a different epistemological background than TM. Both approaches are based on the same facts, assuming the current socio-ecological crisis, but their attitudes are not the same. On one hand, the DA agenda is based on the strong

[13] As it is stated by Marga Mediavilla in her recent Open Letter to Fridays for Future: https://www.15-15-15.org/webzine/2019/12/22/open-letter-to-fridays-for-future-its-time-for-action-but-what-action/.

conviction that short-term collapse is unavoidable and for this reason, resilience should be followed by relinquishment, restoration, and reconciliation (Bendell, 2018). On the other hand, TM embraces a deliberately positive vision, focused on a future plenty of hope (Hopkins, 2019). Which story could be the most appropriate?

The positivism inherent in TM is helpful because it makes our capacities stronger, builds community, reconnects us with nature, boosts resilience, and reduces environmental impacts through bottom-up practices. The power of creativity and imagination to go toward the future we want is probably one of its best tools. Nevertheless, when the political and metabolic implications of reducing our emissions and changing our lifestyles to not worsen the collapse is understood (Bardi, 2017), it becomes apparent that this "Great Turning" (Macy & Johnstone, 2012) is almost unattainable. Which emotional and psychological implications may this awareness cause on our own lives?

DA proposes a main question to guide our reflection on how to navigate our climate tragedy: What could we make peace with to lessen suffering? Do we have to learn to die (Scranton, 2015)? Such questions incorporate the idea of reconciliation with one's death, including any difficulties and regrets in one's life, any anger toward existence itself. That is the pillar of the DA that invites us to make peace with everything and those with whom we have enmity or remorse. And it does so with a particular emphasis on "normalizing" our relationship with death and the dark side of existence. This work can be understood also as aligned with inner transition and the Work That Reconnects, developed by Joanna Macy (Macy & Johnstone, 2012). So maybe in fact both stories and movements are not so different. After all, as Jem Bendell declares, "[o]urs is a time for reconciliation with mortality, nature and each other."[14]

After this review, it appears that social movements must take advantage of windows of opportunity to achieve their goals, whether they are created by the movement itself or as a result of something happening in the world (Speck, 2019: 117). And what is happening now in the world is that the new trends seem to be going to a progress in transition approaches and a renewal replacement through new and young activists.

Moreover, this change in new trends is consistent with the new networked social movements life cycle proposed by Castells (2011) after the

[14] Source: https://jembendell.com/2019/01/09/hope-and-vision-in-the-face-of-collapse-the-4th-r-of-deep-adaptation/.

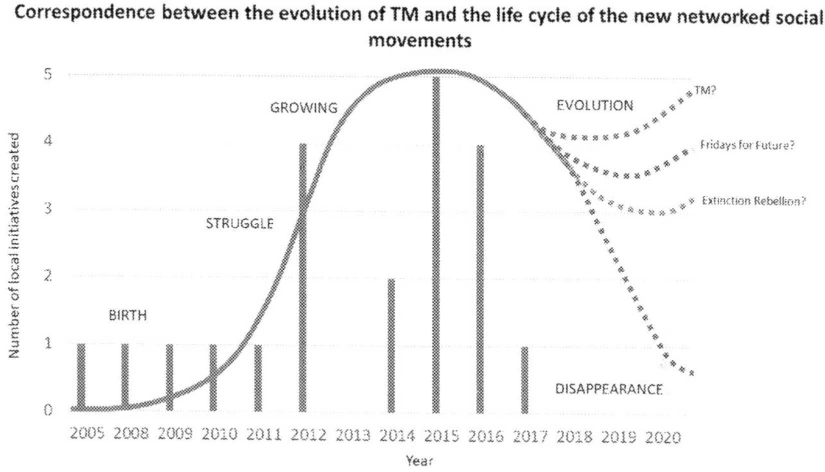

Figure 11.8 Correspondence between the evolution of TM and the life cycle of the new networked social movements. *TM,* Transition Movement. *Red de Transición España. (2020*). El movimiento de transición en el Estado español. Actualización del mapa de iniciativas locales de transición (2014–2017*)*.

systematic study of the new social movements that have emerged in the world since 2008. This life cycle is made up of different stages (birth, struggle, growing, evolution, and disappearance) and seems to show a correspondence between this cycle, the evolution of TM in Spain, and alternative scenarios of the future (Fig. 11.8).

For this reason, it would be appropriate to explore ways of getting closer to and collaborating with these new movements, especially when they recognize municipalities in transition as appropriate alternatives "to forge collective relationships" (Camargo & Martín-Sosa, 2019: 257) and because the long-term organizational needs of emerging climate movements after the celebration of COP25.[15]

[15] At the time of delivery of this chapter, the 25th edition of the Conference of the Parties (COP25) from the United Nations Framework Convention on Climate Change (UNFCCC) was held in Madrid. Curiously, this event has been chaired and organized by the government of Chile, after the initial resignation of Brazil and after the social explosion in this Andean country, which did not guarantee the safety of the participants during the summit. Despite the fact that the alternative social summit gave prominence to the social movements of the Global South, "an abyss can be seen between the umpteenth failure of governments, scientific alarm and the growing mobilization of civil society and environmental organizations" (Santiago-Muíño, 2020).

It is clear that these new trends can influence the future of the TM in Spain. However, this does not represent a disadvantage. Instead, "providing and sharing [different] "worldviews to initiate the transition" from the various situations in the world could constitute factors of extraordinary importance for the essential raising of stakes at all scales (from personal to global) in the face of the threat of civilisational collapse" (Taibo, 2017: 93). Looking for synergies with different ecosocial narratives is precisely an opportunity to boost resilience and regeneration processes, and indeed this is the TM purpose. Community led-action should include networking to catalyze the socio-ecological transformation that our emergent context requires.[16]

11.6 Closing remarks

TM is an example of social innovation and citizen participation that has been able to communicate hope that transformations can produce fuller lives in sustainable societies (Fernández-Casadevante & Piñeiro, 2019: 140). Without the need for a formal ascription to the movement (Zarzuela, 2010), its practices and tools have been assumed by multiple collectives throughout the state.

However, the situation in Spain is currently very complex due to the addition of several concerning factors: levels of inequality among the highest in Europe after the last major recession, political blockade between parties for 4 years, serious territorial crisis in Catalonia, high risk of contagion from the social explosion in Latin American countries, breakdown of the social elevator for younger generations, or the end of the Spanish exception with the entry of far right in the State Parliament. These are factors shaping a future that, according to some authors, is foreseen as a "new normal" based on programmed improvisation, total instability and total adaptability (Niño-Becerra, 2019).

We hope that the outcome of the rethinking process of the TM in Spain will be fruitful. Its original configuration, which has been criticized by some authors for its political vagueness and reluctance to address the

[16] It is already a work in progress by increasing projects, like the Spanish Transition Hub. And at international level, besides Transition Network, for instance, ECOLISE organization is a pioneer European networking concerned about reshaping the future through engaging community initiatives with huge regeneration potential (Penha-Lopes & Henfrey, 2019).

issue of political power (Casal, 2016: 232), may represent an advantage over other approaches on ecosocial transition.

And to face this challenge, a few last ones by Yayo Herrero, who prologued a great book on resilient and exciting practices of the international municipalist movement: "a resistant and creative movement requires, at the same time, belligerence and tenderness, and can only be built by adding efforts (. . .). It is not an easy task and is full of tensions and conflicts. There is much to deconstruct and there are no safe maps to do it, but it is a necessary endeavor, joyful and full of meaning" (Herrero, 2018: 6).

References

Bardi, U. (2017). *The Seneca Effect: Why growth is slow but collapse is rapid*. New York: Springer.

Barrero, L. (2019). *El mundo que nos dejáis*. Barcelona: Destino referentes.

Bendell, J. (2018). *Deep Adaptation: a map for navigating climate tragedy*. Ambleside: IFLAS Source. Available from http://www.lifeworth.com/deepadaptation.pdf.

Camargo, J., & Martín-Sosa, S. (2019). *Manual de lucha contra el cambio climático*. Madrid: Libros en Acción.

Candón, J., & Redondo, D. (2013). "Redes digitales y su papel en la movilización". In M. Cruells, & P. Ibarra (Eds.), *La democracia del futuro. Del 15M a la emergencia de una sociedad civil viva*. Icaria: Barcelona.

Casais, E. (2011). La respuesta a la crisis en la UE: España camino de su "década perdida. *Problemas del Desarrollo*, *42*(166), 37–62.

Casal, M. (2016). *La izquierda ante el colapso de la civilización industrial*. Torrejón de Ardoz: La oveja roja.

Castells, M. (2011). *Redes de indignación y esperanza. Los movimientos sociales en la era de Internet*. Barcelona: Alianza Ensayo.

Chelleri, L., & Del Río, J. (2017). Transition in Spain: A first assessment of dimensions, challenges and opportunities for transition town initiatives. In T. Henfrey, G. Maschkowski, & G. Penha-Lope (Eds.), Resilience, community action and societal transformation. People, place, practice, power, politics and possibility in Transition. East Meon (Hampshire, UK): Permanents Publications.

Del Río, J. (2019). *Guía del movimiento de transición. Cómo transformar tu vida en la ciudad* (2nd ed.). Madrid: Catarata.

Delibes, M. (1979). *Un mundo que agoniza*. Barcelona: Plaza y Janés.

Escorihuela, J. L. (2012). *Camino se hace al andar. Del individuo moderno a la comunidad sostenible*. Samos (Lugo): Nous editorial Source. Available from http://www.elcaminodelelder.org/recursos/CaminoSeHaceAlAndar.pdf.

Fernández, J. (1999). *El ecologismo español*. Madrid: Alianza editorial.

Fernández-Casadevante, J. L., & Piñeiro, C. (2019). Innovación social y participación ciudadana". In C. Álvarez (Ed.), *Informe sobre sostenibilidad en España 2019. Por qué las ciudades son clave en la transición ecológica*. Madrid: Fundación Alternativas.

Fogg, N., & Harrington, P. (2011). Rising mal-employment and the Great Recession: The growing disconnect between recent college graduates and the college labor market". *Continuing Education Review*, *75*, 51–65. Available from https://eric.ed.gov/?id = EJ96780.

Herrero, Y. (2018). Vivimos un momento excepcional de la historia". In M. Junqué, & K. Shea (Eds.), *Ciudades sin miedo. Guía del movimiento municipalista global.* Barcelona: Icaria.

Hopkins, R. (2011). *The transition companion. Making your community more resilient in uncertain time.* Totnes: Green Books.

Hopkins, R. (2019). *From what is to what if: Unleashing the power of imagination to create the future we want.* London: Chelsea Green.

Jiménez-Gómez, I., & Olcina-Alvarado, M. (2017). *Manual de comunicación para una ciudadanía organizada.* Madrid: Libros en acción.

Macy, J., & Johnstone, C. (2012). *Active hope: How to face the mess we're in without going crazy.* Novato (California): New World Library.

Marina, J. A., & Rambaud, J. (2018). *Biografía de la humanidad.* Barcelona: Ariel.

Mella, J. M. (2013). "Hay soluciones a la crisis?". *In Público. Universidad Autónoma de Madrid.* Available from https://blogs.publico.es/econonuestra/2013/06/02/hay-soluciones-a-la-crisis/.

Municipalities in Transition. (2018). *The MiT Framework for the Pilots.* Available from http://municipalitiesintransition.org/wp-content/uploads/sites/3/2019/09/MiT-Framework_version-1.1-April-2018-to-be-published.pdf

Niño-Becerra, S. (2019). (with Vila, N) *El crash. Tercera fase.* Barcelona: Roca editorial.

Penha-Lopes, G., & Henfrey, T. (Eds.), (2019). *Reshaping the Future: How communities are catalysing social, economic and ecological transformation in Europe. The first status report on community-led action on sustainability and climate change.* Brussels: ECOLISE.

Piulats, O. (1992). Teoría y Praxis de la política verde en el Estado español". *Revista de Ecología Política* (4), 139—155, 2nd semester 1992.

Puente, I. (2003). *El comunismo libertario.* Bilbao: Likiniano Elkartea.

Puleo, A. (2017). *What is ecofeminism?". Quaderns de la Mediterrània.* Barcelona: IEMed, n. 25.

Red de Transición España. (2020). *El movimiento de transición en el Estado español. Actualización del mapa de iniciativas locales de transición* (2014—2017).

Ross, M. (2005). *The basics of permaculture design.* London: Chelsea Green.

Ruiz-Aja, L., Pérez, F. M., & Gómez-Pastrana, T. M. (2013). *El descontento social y la generación". 15M: in-conformismo in-ternacionalizado.* Madrid: Editorial Popular.

Santiago-Muíño, E. (2020). *COP25: comprender y superar el enésimo fracaso.* Available from https://www.elsaltodiario.com/tribuna/cop25-comprender-superar-enesimo-fracaso.

Scranton, R. (2015). *Learning to die in the Anthropocene: Reflections on the end of a civilization.* San Francisco: City Light Publishers.

Speck, A. (2019). *Nos organizamos para el cambio social. Pequeño manual de organizing.* Sevilla: Ferrera Ledesma.

Taibo, C. (2017). *En defensa del decrecimiento. Sobre capitalismo, crisis y barbarie* (4th ed.). Madrid: Catarata.

Transition Network. (2020). *Transition Hubs Group: Creating new structures and processes to catalyse and support Transition internationally.* Available from https://transitionnetwork.org/about-the-movement/international-hubs-circle/.

Zarzuela, J. (2010). La comunidad toma las riendas". *El Ecologista* (64), Consulted. Available from http://www.reddetransicion.org/wp-content/uploads/2014/02/La-comunidad-coge-las-riendas_Javier_Zarzuela_2010.pdf.

CHAPTER 12

Resilience and conservation of urban commons: lessons from three community-restored lakes in Bengaluru

Amrita Sen[1,2], C.S. Dechamma[3] and Harini Nagendra[2]
[1]Department of Humanities and Social Sciences, Indian Institute of Technology Kharagpur, West Bengal, India
[2]Azim Premji University, Bangalore, Karnataka, India
[3]Community Engagement Initiative, Azim Premji University, Bangalore, Karnataka, India

12.1 Introduction

This chapter discusses collective and multiactor interventions by local communities in Bengaluru, in conserving urban ecological commons—specifically, urban lakes—which provides a range of services to residents, as well as protecting the overall resilience of the city. Bengaluru, which once had an agrarian ecological landscape nourished by a large network of interconnected rainwater harvesting structures—tanks or lakes, has now grown to a megacity (Nagendra, 2016). Rapid urbanization has been accompanied by conversion of many of these lakes into other forms of land use, with a decline in the functioning of lakes and their surrounding reliant socio-ecological systems. With the import of piped water to the city since the early 20th century, lakes lost much of their perceived relevance for policymakers (Nagendra, 2010). Waste discharge and sewage eventually polluted most of the lakes and choked the overflow channels that connected lakes along a topographic gradient, reducing the flow of water. In recent years, spurred by a resurgent awareness of the importance of lakes, a growing number of civic and community efforts have resulted in lake restoration, in collaboration with the Government.

In this chapter, we discuss a series of community actions to restore lakes as commons in Bengaluru. Drawing on observations from three restored lakes, Kaikondrahalli, Jakkur, and Puttenahalli, we discuss how initial civic attempts to restore the lakes began and how the restored lakes

[ECO]systems of Resilience Practices
DOI: https://doi.org/10.1016/B978-0-12-819198-9.00004-1

became embedded in the lives of the community, as a social–ecological space. Fieldwork for this study had been conducted over an extended period of time, with two of the authors themselves being associated with the restoration and maintenance of one of the lakes for a long time.

Primary interviews were conducted with visitors of the lake along with many of the actors who played a key role in restoring the lakes. We argue that alternative framings of environmental consciousness, which are contemporary and dynamic, are strongly needed to address the growing impacts of environmental degradation in multicultural landscapes like megacities. Such diverse environmental consciousness can lead to a range of concerted actions, especially among communities that are affected by the degradation and decay of urban ecological commons like lakes. We show how, despite being disparate groups, each of these lake restoration collectives shared a common vision of ecological conservation and a shared consciousness that commons need to be restored not solely for their intrinsic ecological value, but also for their role as collective social spaces in increasingly fragmented urban neighborhoods.

The manuscript is divided into three sections, following the introduction. In the next section, we discuss early initiatives on restoring the three lakes: Kaikondrahalli, Jakkur, and Puttenahalli. The section following this includes primary interviews with activists of three lake restoration groups of the three lakes. This section speaks about personal narratives—the relevance of nature in the everyday lives of the group functionaries and how that sparked concerns about environmental sustainability and, eventually, restoration of the sewage-chocked lakes. Finally, we describe ways in which restored lakes like Kaikondrahalli has become nodes of environmental placemaking—a term which we use to explain the emergent, dynamic processes by which diverse social communities build meanings and attachments to places of nature, shaping stewardship and collective action for conservation (Sen and Nagendra, forthcoming). Instances of environmental placemaking are demonstrated using primary field interviews from Kaikondrahalli Lake.

12.2 Community restoration of three lakes in Bengaluru: Kaikondrahalli, Jakkur, and Puttenahalli

Jakkur Lake, located in north Bengaluru, covers 55 ha with three inlets and two outlets. The lake was restored between 2009 and 2011. Before this, the lake faced a number of challenges, including land conversion, soil mining, and sewage pollution. Restoration efforts were conducted by "Jalaposhan," a citizens group led by a well-known local medical doctor, Dr. Annapurna Kamath, in collaboration with government agencies. The Bangalore Water Supply and Sewage Board (BWSSB) established a sewage treatment plant (STP) capable of processing 10 million liters per day, just upstream of the lake. A wetland and algal pond were created adjacent to the STP. Treated water from the STP is taken to the wetland, which helps to absorb the excess nitrogen and phosphorus present in the treated water (The Indian Express, 2018). This water then enters the lake (Fig. 12.1).

Tree plantation was conducted around the lake, to increase biodiversity and provide habitats for birds, butterflies, and bees (Lekshmi Priya, 2018). A number of experts were consulted, who volunteered their expertise and inputs. The lake today houses around 300 species of plants, including red-listed plants and native medicinal plants and herbs, used by many visitors as herbal home remedies for cough, cold, diabetes, and other ailments.[1]

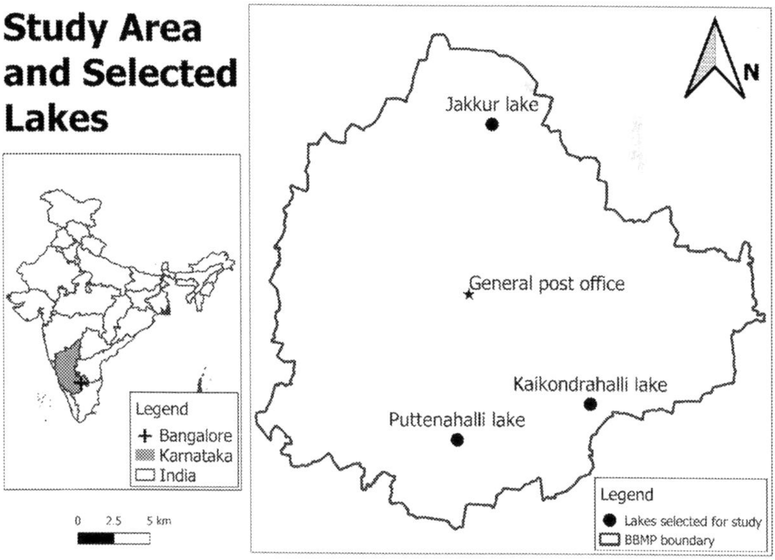

Figure 12.1 Map showing the location of the three lakes. *Credit: Map prepared by Enakshi Bhar.*

[1] https://bangaloremirror.indiatimes.com/bangalore/cover-story/it-was-the-right-dose-of-herbs-that-brought-this-lake-back-to-life/articleshow/66339102.cms. Accessed on 13/01/2020

There is a separate butterfly garden, as well as specific enclosures for nectar-intensive plants and tree species, along with a permaculture garden for community gardening.[2] Tree plantation drives and tree festivals or "Vanamahotsava" have been carried out with active participation of local residents. The restoration of the lake is yet another example of the success of collective action in ecosystem restoration in the urban context.

However, the lake is not free of pollution. Recent findings indicate that stormwater drains near the lake are discharging untreated sewage water into the lake. The recent Yelahanka Power Plant also poses a significant threat to lake, as it utilizes a lot of the treated STP water that should go to the lake.

Kaikondrahalli Lake that was once a thriving ecosystem underwent a lot of changes within two decades. Today, it stands as a steady example of community power, relentless contributions, and a helping hand to many communities in and around to restore many other lakes. Kaikondrahalli Lake is situated in the southeast part of Bengaluru. Its eastern boundary shares the boundary of Sarjapur road giving each passerby a slight peak of the lake. The lake is spread around 48 acres with a 1.9-km perimeter and with a wide variety of flora, fauna, insects, reptiles, butterflies, and birds. More than 50 species of birds mostly native and some migratory were spotted by some of the bird lovers around the lake which is seen to be on the rise each year.

Like many lakes in Bengaluru, Kaikondrahalli Lake also played an essential role to the community by providing many fold ecosystem services. The community around the lake also maintained the lake to its highest capability to derive services out of the lake which involved fishing, grazing, agriculture, de-silting, firewood collection, and many cultural activities protecting the lake. This kept both the lake and the activities around it, hand in hand until it hit a bump. The lake in the mid-1990s was neglected as the need for agriculture decreased and water from the lake wasn't used for drinking purposes by both communities around the lake and the panchayat. The lake's situation worsened in the early 2000s because of the developments that took place around the lake. Real estate value hit the sky and high-rise buildings were built near the lake. This became one of the sole reasons for the lake to degrade and the agriculture lands and surrounding wetlands were encroached for real estate purposes.

[2] https://www.thebetterindia.com/163949/jakkur-bengaluru-lake-pollution-cleaning-news/. Accessed on 13/01/2020

Kaikondrahalli Lake then became a cesspool and a dumping ground—construction waste, domestic waste, poultry waste, medical waste were dumped illegally. The rainwater inlets and the natural kaluves were choked. In the last stage, it saw a lot of sewage entering into the lake, making the lake unfit for any purpose and the killing the entire ecosystem.

It all started with a very passionate resident Priya Ramasubban, a documentary filmmaker by profession who had then moved into the locality. It was when she took walk around the lake, she found a group of colorful dragonflies lying around in the corner, she realized that there is life thriving in the dead lake. Fortunately, in the same year 2007 Bruhat Bengaluru Mahanagara Palike (BBMP) had issued a public notice to restore some of the lakes in Bengaluru and one being Kaikondrahalli Lake. She saw this as an opportunity for the community to involve in the restoration alongside the municipal corporation. Soon like-minded individuals Ramesh Sivaram a social activist, Rajesh Rao a resident of Ambalipura, and David Lewis a senior citizen and a resident of Kaikondrahalli joined hands with her and formed a resident group and met experts like ecologists, ornithologists, herpetologists, civil engineers, and also some local villagers to understand diverse inputs for an inclusive restoration. Later they met the then Chief Engineer of lakes Mr. Satish B.V who headed the Lakes Department of BBMP and who saw potential in community-driven rejuvenation. The Detailed Project Report that was already prepared for the restoration was shown to the group. With the help of experts, they could negotiate with the BBMP to redesign the rejuvenation, focusing on social and ecological aspects that catered to the local needs (Nagendra, 2016: 4).

Ideas such as having washroom facilities, an open amphitheater, a Kalyani which saved the lake water from getting polluted during Ganesha and Durga festivals were included in the design. A design that had a native approach keeping the lake as the center of the focus and bringing back all the traditional livelihood that once took place was thought about.

These ideas were welcomed by the Chief Engineer and a team of sincere officials was formed consisting of Assistant Conservator of Forest Mr. Jagganath Rao and Assistant Engineer Ms. Shilpa Karthik. The rejuvenation took place in two phases. First, it started with mapping the lake boundary, marking rainwater inlets or kaluves, and fencing. Then they started with de-silting the lake, laying sewer pipes, island for birds and reptiles, and forming the lake bund. In the year 2009–11 first phase of the rejuvenation was completed. After the first phase of rejuvenation the local resident group having Priya Ramasubban, David Lewis, Ramesh Sivaram, and Rajesh Rao formed a trust, Mahadevpura Parisara Samrakshane Mattu Abhivrudhi Samiti (MAPSAS) (Nagendra, 2016: 5). The trust entered a tripartite agreement by signing a Memorandum of Understanding (MoU) with the BBMP and United way a funding agency to maintain Kaikondrahalli Lake. MAPSAS is given the role of primary custodians of the lake to monitor the daily maintenance of the lake, BBMP in charge of the infrastructural repair works, and United way for the financial assistance.

In the second phase of rejuvenation which took place between 2011 and 2012, walking path, tracks for physically challenged, fixing curb stones, and toilets were made. After this phase, Kaikondrahalli Lake attracted a lot of visitors. During the monsoon, rainwater entered the lake and soon a wide variety of birds, reptiles, insects, butterflies made the lake their habitat. Social and ecological aspects of the lake improved. With good water level, tenders for fishing were called out by the Government of Karnataka's fisheries department. MAPSAS also engages with the local community to keep the relationship in tandem. They encourage traditional cultural and religious practices that have passed on for centuries to be practiced at the lake. It could be praying to the deities of the lake to the trees and water in the lake. They also support livelihoods for low-income communities living around the lake by creating creative entrepreneurship practices like the women from the community make use of the weeds from the lake that are grown abundantly into handicrafts. Each year they implement new projects at the lake such as leaf-composting units, butterfly gardens that attract 15 different species of butterflies, play area for the kids, herb garden, sewage diversion, and wastewater treatment plants. Looking at the efforts of the local community in maintaining the lake, BBMP involves them in the restoration of the other lakes in the chain. Now MAPSAS is the primary custodian for four lakes, Kasavanahalli, Soule Kere, Lower Ambalipura, and Kaikondrahalli. Apart from this, they act as advisory bodies for other lake groups giving them technical assistance.

Since MAPSAS was a small core group of residents, they wanted to expand by involving people from the community. Keeping this in mind, they conducted activities at the lake during the weekends for kids and adults. Art for children and adults, like art for kids and de-silting for elders. They also conducted an annual event called "Kere Habba" (lake festival) to celebrate the lake's existence with a lot of environment-related activities ranging from native food stalls to environmental solutions, for all age groups. This attracted about 3000 visitors in the first year and has been increasing each year. This event generates funds for the yearly maintenance of the lake. Now the trust consists of 9 trustees to 400–500 volunteers working for the benefit of the lake. Each year the number of trustees and the volunteers increase. Accounting an average footfall of 400 people per day, the lake attracts people from different backgrounds such as bird watchers, photographers, senior citizens, runners, cyclists, grazers, garment workers who walk through the lake to bypass the polluted jam-packed and unsafe roads.

For the upkeep of the lake the trust receives money from individual donors, from corporate groups, and fund-raising events such as "Kere Habba." The money is utilized to fund guards, gardeners, and other workers and to implement small projects at the lake.

The restoration of Kaikondrahalli Lake provides an example of the potential of urban collective action in ensuring resilient, equitable, and environmentally sustainable conservation and restoration, which can in turn contribute to a shared ecological consciousness. Post restoration, the lake witnessed significant physical and ecological transformations, including an increase in the number and diversity of birds, frogs, butterflies, and snakes. As one of the few remaining open spaces in the neighborhood, the lake now attracts a large group of visitors. A number of different community events and festivals are organized at the lake. The most popular among these is the *Kere Habba*, or annual lake festival, held every year in January or February. There are programs, cultural events, panels, and popular demonstrations by children and young visitors on this occasion, organized around the theme of nature and sustainability (Nagendra, 2016, 6). MAPSAS, a local trust created by residents who had worked toward the restoration drive, still continues to manage the lake, helping ensure that the lake remains in good condition.

Despite sustained efforts, the trust continues to battle a number of challenges in protecting the lake. Over the years, the condition of the lakes has started to deteriorate, due to unchecked construction on the wetland and valleys that drain into the lake, and the dumping of waste material and effluents near the lake bed (NITI Aayog, 2015, 87; Nagendra, 2016, 9). Political contestation has posed a challenge. A municipality ban on traditional livelihood practices like cattle washing has exacerbated social inequity.

Irrespective of these challenges, Kaikondrahalli Lake offers a very inspiring and a positive story. With responsible and passionate individuals giving light to the problem in the direct direction, involving larger community, and understating the importance of the resource catering enormous benefits to the community around, we will be able to see more and more common pool resources thriving. Because of the relentless participation of the community in the restoration process, Kaikondrahalli Lake today stands as a "model lake" for India named by the NITI Aayog.

Another well-known restored lake is Puttenahalli Lake. Puttenahalli Lake is situated in the 7th phase of J.P Nagar in the South of Bengaluru. The lake is a small lake, spread across 13 acres 25 gunthas with a walking perimeter of 1.1 km. Like most lakes in Bengaluru for centuries, Puttenahalli Lake was once a thriving ecosystem which catered to the livelihoods of the people living around it. The lake's condition deteriorated in the early 2000s because of the rapid urbanization around the lake. The lake was converted into a garbage dump and became a breeding site for mosquitoes.

It was in the year 2006 when Usha Rajagopalan, a local writer, moved into the locality. While her house faced the lake, each day as she looked at the lake, its condition deteriorated and the dumping of debris increased, with the increase in water problems during summers, she thought Puttenahalli Lake should be restored and turned into a catchment like it always did. Usha Rajagopalan and three of her neighbors started a campaign "Save the Lake" (Rajagopalan, 2017). Efforts were put to get the attention of the elected representative of the respective constituency and media. Puttenahalli Lake was added to the list of lakes that BBMP planned to rejuvenate that particular year. The first phase of rejuvenation started in the year 2009 with marking the boundary and fencing. Usha Rajagopalan and the team conducted activities on Earth day in the neighborhood to get more attention and include more people from the locality. They formed the Puttenahalli Neighbourhood Lake Improvement Trust (PNLIT) in the year 2010, to maintain Puttenahalli Lake and to oversee problems in the neighborhood. At the same time the BBMP started full-fledged rejuvenation of the lake by making walkways, de-silting, and connecting the kaluves. PNLIT started hosting activities such as planting, clean-up drive in the lake to attract volunteers who could bring in funds for the maintenance of the lake. In the year 2011 they signed an MoU with the BBMP to become the primary custodians of the lake. With the trust coming into form they explored new sources of water apart from direct rainwater and water from the upstream lake. One such idea was to direct run-off water from the neighboring apartment and letting excess treated water into the lake with the Karnataka State Pollution Control Board's standards. The water quality was then sent to BMS College of Engineering for quality check. This way they could get the students in the neighboring college involved in scientific studies of the neighboring resource.

The trust partnered with different non governmental organizations (NGOs) and organizations. The lake receives a substantial amount of water from the STP connected to neighboring apartments. The need for having good wetlands was a priority, because of the N- and P-enriched wastewater entering the lake. A couple of floating islands with biofilters were additionally installed for this purpose. With the help of grazers and fisherfolk, the lake is well maintained. Weeds are removed timely by the grazers and BBMP workers and the fishermen keep a check on the water inlets and water quality. They host numerous activities from planting trees, preparing leaf-composting units at the lake, bird survey to storytelling, music, and art.

Most of the funds for the maintenance of the lake are collected from individual donors and donation drives. Some projects at the lake are implemented by corporate support. This has been a challenge for PNLIT to build corpus and they have been looking out for a steady support. They too face challenges like other lake community such as encroachments and sewage breach. But they have been able to get timely help from the BBMP and vigilant residents.

Puttenahalli Lake has received a number of awards, including the Namma Bengaluru Award, 2012, and My Livable City Award, 2017. For the trust the "biggest award would be for Puttenahalli Lake to thrive" (Rajagopalan, 2017). The success story of the community led restoration of Puttenahalli Lake also provides inspiration for 30 other citizen groups that were formed after PNLIT. Although they face challenges at each stage, the trust says there is no looking back for them. Puttenahalli Lake is home to 103 different species of birds, insects, butterflies, and a wide variety of flora and fauna.

Jakkur Lake is another example of a community participatory model of restoration. Jakkur Lake compared to Kaikondrahalli and Puttenahalli Lakes is a bigger lake, spreads across 160 acres, forms an important and larger lake in the lake chain, and is located in the northeastern part of Bengaluru. The main source of water to the lake is the treated water from

STP that is set up by the BWSSB and rainwater and water from the upstream Yelahanka Lake (Jamwal, 2017).

Jakkur Lake, which once supported expansive ecosystem and livelihoods that depended on it, was hit by the urbanization and developmental work with the rise in the economic growth in the late 1990s. The lake for centuries was the main source of water for the village. Water was used for drinking, washing, fishing, irrigation, for cattle and other animals in the village. Jakkur Lake was also used by more than 50 dhobi families (The Indian Express, 2018). Changes in land use patterns around the lake resulted in encroachments of the lake land, which was also used as a dumping ground. Waste from the apartment complexes, poultry waste, and other debris along with domestic sewage made its way to the lake further polluting the ecosystem. In 2005 the Bangalore Development Authority constructed a fence around the lake to provide protection. In 2004 BWSSB set up a 10 million litres per day (MLD) STP with an upflow anaerobic sludge blanket.

In the year 2008 the first phase of restoration work commenced. Desilting, construction of bunds, island, and walking path and a small water tank were constructed in the first phase. About 9 acres of artificial wetlands were created with algae ponds for further treatment of water before letting it into the lake. About 8–9 MLD of treated water was released into the lake. This saw increase in the level of water in the lake and good quality water stored in the lake. Every fortnight the water quality is tested in the in-house laboratory. Although the community is vigilant of the quality of water entering the lake by monitoring the test reports diligently. Alongside Satya Foundation a charitable trust, working in the nature conservation, involved in the Jakkur Lake restoration to protect and conserve the lake through their project called "JALA POSHAN"[1]. In 2014 Jalaposhan became a trust with Dr. Annapurna Kamath managing the trust. Jalaposhan was closely involved in the restoration process, building a community network and helping the government by taking over the monitoring and maintenance of the lake. They brought in innovative approaches to lake management by allowing fishing, grazing, and creating an herb garden that now has more than 300 varieties of herbs, community farming, and promoting women's self-help groups by providing jobs for the women from the locality. By building networks with different departments of the government, the trust has been able to access resources such as getting native fruits and flowering plants from the Horticulture Department. Jalaposhan hosts activities and events at the lake every week.

These range from tree planting drives, to nature walks for the students and music programs for residents. This has seen an increase in the footfall of not just the visitors but also volunteers who are willing to participate in different activities. Funding for the lake maintenance largely comes from community donations, fund-raising events such as Kere Habba and corporate grants. Like other lake groups, this group also finds it challenging to maintain the lake without a steady inflow of funds.

Sewage from nearby residences continues to enter the lake. This has been a constant challenge for the community in terms of figuring a long-term solution. Apart from this the government plans to set up a power plant at Yelahanka by the Karnataka Power Corporation Ltd. This will put the pressure on the lake as 10 MLD of the water will be diverted to the power plant. Since the major source of water to the lake is treated water, this will affect the water levels of the lake. Amidst these challenges, Jakkur Lake is now a thriving rich ecosystem with multistakeholder engagement.

All three lakes, restored by the efforts of local communities in collaboration with government agencies, have helped to regenerate biodiverse, functioning ecosystems that are also social–ecological spaces. The lakes are frequented by a range of people, who have different reasons for visiting the lakes and multiple ways of identifying with the lake landscape. It is due to these bonds established with the place that integrative and collective attempts of restoration could be designed and implemented.

12.3 Restored lakes as nodes of environmental placemaking

In the global South, an accelerated drive toward urbanization has led to widespread immense degradation of local environment and ecology. Urban environmentalism seems to be on the rise in many cities, in response to these challenges. Sheppard (2006, 300) broadly defined urban environment as the "air, water, soil and green, grey and brown infrastructure of city centre areas, suburban areas and fringe areas, including urban flora, urban non-human fauna and human beings, as well as the habitats of each" and urban environmentalism as "the collective effort to maintain, safeguard and protect the complex urban environment and urban environmental function."

We locate "environmental placemaking" as a precursor to urban environmentalism—a process that helps us understand the emergent, dynamic processes by which diverse social communities build meanings and attachments to places of nature, shaping stewardship, and collective action for conservation (Sen & Nagendra, 2018). In fast changing global cities like Bengaluru, environmentalism is not only shaped by deep-rooted intrinsic cultural affinities for nature, and long-standing livelihood dependencies on the ecosystems, but through a shared sense of place-identity and networks of place-based collective action (ibid.). The term placemaking has specific epistemic relevance—one which underscores presence of collective and interspatial practices in sustaining a region as a "place" (Massey 1994; Jonas, 2011). Environmental placemaking transcends the idea that standard and representative practices are fundamentally amenable to nature conservation. On the contrary, it enables an analysis of the dynamism of environmental stewardship concerns in multicultural postcolonial landscapes like Indian cities, which are going through widespread and rapid change.

Field interviews with lake visitors documented a number of instances of environmental placemaking. Many senior citizens, who meet daily at the lakes, say that they came to Bengaluru after retirement to see their children, who work in the city. The lake is a place essential for them, since it provides them with a sense of companionship. "This place is our rescue from boredom," says one, who feels that his son and grandchildren being busy with work and studies, the lake provides a much-needed respite. One of them, from a nearby city—Mangalore—said that he had always been connected with nature and thus he had also actively participated in the restoration of the lake. He had raised funds for planting different kinds of trees. He feels good about the fact that he could participate in the restoration. He often comes to the lake with his pets now. He feels that his pets enjoy being in nature and prefer the lake to being taken for a walk on the busy, traffic-filled streets. A man who lived in a nearby apartment said that he came to the lake every evening with his 4-year-old son. Originally from Ahmedabad, he mentioned that there was no shortage of open green spaces in his city when he was growing up. But he feels depressed to observe how playing in fields is now replaced by playing with indoor video games, play stations, and smartphones. "My sole intension of bringing my child here is to introduce him to nature and provide him with a healthy childhood," he says. He also mentioned that he had introduced to him the colors and names of animals, plants, butterflies, and flowers that grow here. "For me, making my child aware of the beautiful side of the world is very important. Kaikondrahalli lake has become an integral part of his learning place," he said.

A 37-year-old female, who is from Nagpur, said that she quit a stressful job after she came to visit Kaikondrahalli Lake once with her sister, who stays in a nearby apartment. Her sister brought her here one evening after leaving her children to the tuition class. Sitting close to a bird-nesting area and watching different birds flying around, they discussed stories of restoring the Kaikondrahalli Lake. She was mentioning how amused she was when she heard about the restoration of the lake.

> *The hidden spirit of being a wilderness lover tickled in me. I started dreaming of implementing this successful restoration process in a jogger's park around a pond just behind my house in Nagpur. The dream dimmed soon as I was back in the room. But as I slept at night, I thought, why not I go there in the morning to enjoy the solitude? It would be a challenge, as my biological clock is already warped because I am a night shift IT employee... The morning freshness reminded me of my dream. I started looking for people who can help me executing the restoration idea in my place. Today we are a team of 7 members, almost near our goal of restoring the jogger's park. After a lot of political and social challenges, the goal is now near....*
>
> *This lake and the story of restoration will be an inspiration throughout my life. I believe this lake gave me a purpose for life and also the courage to quit my job for something better.*

Some of the vendors, who sell lemonades and coconut water outside the Kaikondrahalli Lake boundary, mention that the trees provide shade under which they can comfortably stand and sell, especially in the evenings and afternoons. A lot of visitors come to jog and exercise—after the rounds, many of them become tired and come to have coconut water. Kaikondrahalli Lake, being a popular place, helps the vendors in getting customers.

These unique connections, forged with the lakes, collectively define the idea of environmental placemaking. The lakes to these people are not a mere physical natural setting, perceived as a recreational site, but rather a landscape toward which they have a strong sense of belongingness, acted out through willingness in community activities for lake rejuvenation, among others. Such inclusivity of different actors in the process of restoring an urban common lends a scope for understanding placemaking as a form of environmentalism in a city. Simultaneously it renders a broader understanding toward the multiscalar socio-ecological linkages that shape urban ecology in place-based contexts. It is fascinating to observe how such natural spaces within the boundaries of a megacity like Bengaluru are imbibed within a range of social relations, which, although disparate, are collectively instrumental in making the lakes a part of the urban commons and increasing the resilience of the city.

12.4 Conclusions

This chapter discusses a unique idea of resilience in shaping ideologies of environmental conservation. Drawing on the collective attempts of different urban communities in restoring their individual neighborhood lakes—Kaikondrahalli, Jakkur, and Puttenahalli, and instances of environmental placemaking, we make a case to understand ecological resilience as a significant component of multicultural landscapes. In restoring the three lakes through a collaborative initiative of the communities, state, and the civic groups of Bengaluru, we discerned a distinctive meaning of the idea of resilience—one that can be shaped through overlapping and disparate, yet new and innovative ways. Such dynamicity in collective community actions and the ideology of resilience framed therein helps explore contemporary methods of environmental conservation—ones that are not standardized or representative but fostered through emerging socio-ecological linkages. A newer idea of ecological resilience helps us understand that nature conservation is a reasonably dynamic and open-ended process, which cannot be explained by any unilinear and representative environmental narrative. The chapter also makes a case to understand how restored urban ecological commons can be defined and imagined as a "place" by its neighborhood communities. The diversities of social–ecological collaborations visible in environmental placemaking help understand how places can be dynamic entities and repositories of collective meanings and practices (Jones and Cloke, 2002 in Sen & Nagendra, 2019). Environmental placemaking helps us understand that urban communities who have heterogeneous forms of associations to ecological commons and diverse histories of belongingness can also contribute toward the process of environmental conservation. The chapter identifies significant scope for future research on ecological resilience, utilizing observations from community conservation initiatives, which are contemporary, adaptive, and robust.

Acknowledgments

Sections on the restoration of Kaikondrahalli Lake are drawn from "Restoration of the Kaikondrahalli Lake in Bangalore: Forging a new urban common," by Harini Nagendra, published by Kalpavriksh, Pune, in 2016. The authors gratefully acknowledge Azim Premji University for funding this research. The authors would like to thank B. Manjunatha and Varsha Bhaskaran for their assistance with fieldwork. Enakshi Bhar is gratefully acknowledged for her help with the map.

References

Jamwal, P. (2017). Fate of nutrients in human dominated ecosystems. *Resonance*, *22*(3), 279–290.

Jonas, A. E. G. (2011). Region and place: regionalism in question. *Progress in Human Geography*, *36*(2), 263–272.

Jones, O., & Cloke, P. (2002). *Tree cultures: The place of trees and tree in their place.* New York: Oxford University Press.

Lekshmi Priya, S. 2018. Citizens turn Bengaluru's Jakkur Lake into eco-zone with these herbs and rare trees. Available at https://www.thebetterindia.com/163949/jakkur-bengaluru-lake-pollution-cleaning-news/.

Massey, D. (1994). *Space, Place and Gender*. Minnesota: University of Minnesota Press.

Nagendra, H. (2010). Maps, lakes and citizens. *Seminar*, *163*, 19–23.

Nagendra, H. (2016). *Restoration of the Kaikondrahalli Lake in Bangalore: Forging a new urban common*. Pune, Maharashtra: Kalpavriksh.

NITI Aayog. (2015). *Lake restoration: two successful models of lake restoration in Rajasthan (Mansagar) and Karnataka (Kaikondrahalli)*. Government of India.

Rajagopalan, U. 2017. How citizens came together to save Bengaluru's Puttenahalli Lake. Available at http://www.earthamag.org/stories/2017/4/3/how-citizens-came-together-to-save-bengalurus-puttenahalli-lake.

Sen, A., & Nagendra, H. (2018). Multiple constructions of environmental placemaking around the urban commons in Bengaluru, India: some empirical examples. *Indian Anthropologist*, *48*(2), 39–52.

Sen, A., & Nagendra, H. (2019). The role of environmental placemaking in shaping contemporary environmentalism and understanding land change. *Journal of Land Use Science*, *14*(4–6), 410–424.

Sheppard, J. W. (2006). The paradox of urban environmentalism: problem and possibility. *Ethics, Place and Environment*, *9*(3), 299–315.

The Indian Express. 2018. Cities at crossroads: who pays to save this lake. The Indian Express, Available at https://indianexpress.com/article/opinion/columns/jakkur-lake-bengaluru-water-pollution-wastewater-management-5196201//

CHAPTER 13

Beyond formal urban policies: resilient periurban alternative practices in the Amazon

Ana Cláudia Duarte Cardoso[1], Thales Barroso Miranda[2], José Júlio Ferreira Lima[1] and Elis Tayná Almeida Brazil[3]
[1]PGITEC, Cidade Universitária, Universidade Federal do Pará, Belém, Brazil
[2]Secretariat of Environment of the City of Belém, Belém, Brazil
[3]Ateliê de Arquitetura e Urbanismo, Cidade Universitária, Universidade Federal do Pará, Belém, Brazil

13.1 Introduction

Over the 20th century, at the Global North, ecological approaches have placed physical characteristics of urban sites as an important issue, providing methodologies to superimpose and match sites' geological features and the anthropogenic interventions (Forman, 2014; Rees, 1992; McHarg, 1969). Cities also were acknowledged as socially produced (Lefebvre, 1991), and in a long-term perspective, it was clear that the relationship between society and nature has been something strongly defined by culture, values, and priorities, according to the availability of economic resources and technologies (Hagan, 2015).

Culture is also placed as a foundation to urbanism by the Italian School of urban morphology, which contraposes the knowledge on how to produce settlements, as an adaptive inheritance, which is accumulated and transferred from one to the next generations, to the formal training and technical education of planners and architects. The school calls these positions as spontaneous (learned from culture) and critical (learned from education) consciences, respectively (Cannigia & Maffei, 1979). This differentiation offers a new perspective to the tension created between culture and economy by market-oriented (capitalist) urbanization, which is reflected in mismatches between outdated formal professional practices and expectations of making urbanization sustainable, following the contemporary acknowledgment that community commitment rooted in culture is important to maintain and enhance the provision of primary

[ECO]systems of Resilience Practices
DOI: https://doi.org/10.1016/B978-0-12-819198-9.00018-1

natural resources (soil, water, biodiversity), among other features (Swann & Deslatte, 2019).

Nevertheless, capitalism has reached parts of the planet differently, Latin America, as part of the South Global, has been integrated as supplier of commodities, and since colonization, it was established that colonial metropolises should follow European urbanization standards, due to the wealth raised by cycles of forest products export, without truly understanding the basis of their wealth—the primary natural resources, especially water and biodiversity, and their difference from industrial poles. Although a discussion on trajectories of development (Costa, 2019) is outside the scope of this chapter, it is important to understand that Amazon region was placed as a periphery of the world system, where rural and wild environments were historically devalued in comparison to cities and deforest "productive" areas, and where colonization was economic but also epistemological (see Fig. 13.1).

As colonizers were selective, they have fostered urbanization in the areas of interest and commanded the production of cities with administrative functions (Becker, 2013) . However, in practice, a coexistence of rationalities and cultures was always established within the two Brazilian

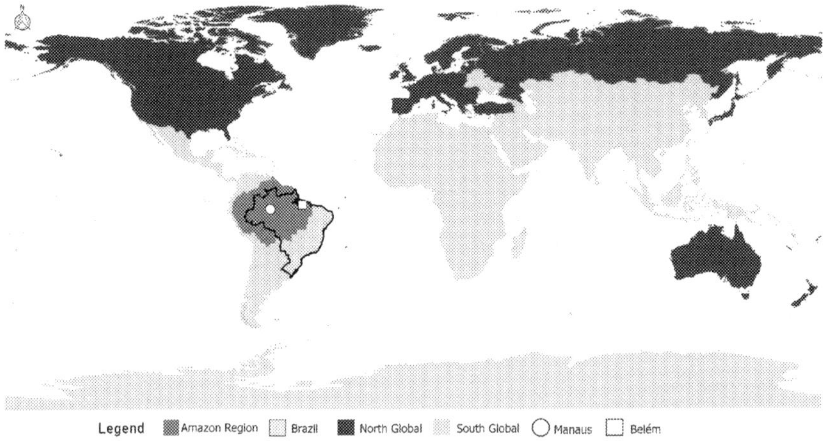

Figure 13.1 Location of Amazon in relation to the world system, Brazilian Amazon, and its two metropolises—Belém and Manaus are highlighted. *Credit: RAISG (2018). Delimitation of the Amazon. São Paulo: RAISG, 2018. [Digital cartography, shapefile format]; IBGE (2018). Delimitation of Brazilian cities. Rio de Janeiro: IBGE, 2018. [Digital cartography, shapefile format]; Thematic Mapping (2009). Delimitation of countries. Great Britain: Thematic Mapping, 2009. [Digital cartography, shapefile format]. Elaboration: Thales Miranda.*

Amazonian metropolises—Belém and Manaus, to the extent that elites used to live in the planned space, and regional peasants who migrated to the cities could use their knowledge on how to occupy floodplains or water and relate to internal rivers.

Under this context, origin defines class and insertion within either formally or informally produced areas, with endogenous occupation[1] listed under the latter. After the 1960s' economic and road integrations, the dream of modernization was reinforced, worsening the conflict between market-led urbanization and availability of primary resources, despite their importance for the livelihoods of the population kept outside the formal market, and their influence on other cities and human settlements located in the Amazon, a peculiar peripheral Global South context with a planetary environmental appeal.

Conversely, according to archeological findings (Heckenberger et al., 2008; Neves, 2006), Amazon forest had an expressive population of 10 million inhabitants before colonization. Indeed the forest has an anthropogenic origin, it is the result of a virtuous relationship between human agency and natural processes. Despite all spatial transformations caused by 20th-century logistics, the same pattern of small size and scattered settlements was maintained by peasants across the last two centuries and has been reedited in periurban areas, since the claim of rural areas by mechanized activities caused the migration of peasants to the outskirts of regional metropolises.

Against this backcloth, this chapter approaches how the epistemological colonization in Northern Brazil was enforced by official regulations and plans, which were market-oriented biased, to rise the occurrence of a more resilient pattern of occupation, that can provide insights on how to build a more endogenous planning agenda. The chapter argues that patterns of urban expansion are related to outcomes of local policies and examines resilient practices of resistance in periurban settlements of Belém metropolitan region that provide means for local livelihoods while preventing built densification and soil impermeability, following the ancient matrix of production from biodiversity. Remote-sensing techniques were used to clarify patterns of urban expansion, to assess fragmentation and occupation of urban fabric, as well as green coverage over a 34-year period of studies.

[1] Endogenous occupation here means the one developed locally, according to local knowledge, resources, and technological availability and with respect to geomorphology.

13.2 Mismatches of peripheral urbanization in Brazil

Brazilian urbanization was actually launched after the 1950s, period of industrialization and provision of logistics to integrate all regions of Brazil, known as "developmentalism." These transformations occurred during the military dictatorship and were fast and oriented by productive sector and market interests. Territorial concerns were left in the backstage due to the big availability of land, rural areas were directed to agrarian settlements, agribusiness, and other forms of natural resources extractivism (dams, mining, timber exploitation), while urban issues were tackled by mass housing production not affordable to lower income groups. Since then, social movements have fought for urban reform, claiming the right to serviced urban land. This claim has resulted in the inclusion of urban reform as a topic of Constitution implemented in 1988, and in the approval of the City Statute (Federal Law no. 10257/2001), very focused on the establishment of private properties' social function.

Meanwhile, the literature from the 1980s was very sharp in explaining how the production of the city became a source of accumulation (Farret, 1985; Gonzales, 1985; Singer, 1982); and it is still suitable to explain how contemporary fractions of capital operate in Brazilian cities (Cano, 2011). It was reinforced after the withdrawal of public sector occurred since the 2016 coup. This literature is still able to describe what happens within cities, despite the advances produced during the existence of Cities Ministry (2003–18), with the due updates about the influence of global financial markets and the growing awareness about outcomes of bad management of environmental features (Ferreira, 2007; Schultzer, 2014).

The aggregation of new territories still is the easiest, cheapest, and most comfortable way—from real-estate sector perspective—to expand the city. In less dynamic economies, it is an incremental process, which is often short sighted in relation to geomorphological characteristics. The easier it is to access financial capital , the stronger the action of real-estate sector toward site homogenization—through landfill or channeling of rivers, dismantling of hills, loss of water sources, deforestation, etc. This is due to the current separation between fields of knowledge that are in charge of the disasters caused by environmental degradation (especially floods) and of territorial planning. In both directions, there is no acknowledgment that to foster the deep-rooted resilience[2] that was developed in

[2] For a broader definition of resilience, see Meerow et al. (2016).

the past (by indigenous communities able to respect natural cycles and capacity to adapt to gradual changes), would be an effective way to prevent vulnerability to climate change and extreme events.[3]

The change toward private appropriation of land parcels and the possibility of creating large properties turned transactions easier, because of the reduced number of parts involved, in a way that large properties or environmentally protected areas became reserves of value to be exploited in the coming future. In a peripheral region known as the Amazon, abundant in natural resources, water, biodiversity, and soil integrity are still taken for granted, and their disappearance is not counted economically, neither is the loss of the knowledge about how to balance human presence in such landscape.

Newly urban turned land parcels that incorporate rural land are designated to new land uses that can pay higher interests than the former rural uses would pay. Land prices are still determined by the expansion of the urban fabric and of land use demands, according to the real-estate market productivity, what is essentially speculative movement. There is no direct correlation between the cost of production of the land and its price. When urbanization is determined by market, the simple control (or private property) of land delivers surplus to its controllers; the same occurs when there is a monopoly of access to land which means to make feasible some activities within the city or to provide access to amenities (Farret, 1985; Singer, 1982).

Conversely, inside this logic, the totality of urban services has its provision irradiated from the urban centers toward the peripheries, including mobility costs as a strong determinant of housing location. Cities within a peripheral context do not offer locations to the poorest groups, since private property is the basic entitlement to access a good location or served urban land, and there is no means to such groups to meet the minimum income requirements to access available land. This fact generates restructuring and new centralities and continuously directs the poorest for the areas outside the market that either might have been excluded from the market by private ownership control or are environmentally fragile, stimulating the production of informal settlements and the uncontrolled built density increases on them (Maricato, 2000; Rolnik, 2018). As the Brazilian Amazon is mostly formed by the Amazon River floodplain and

[3] In Amazon extreme, events have been related to the increase of intensity of droughts and floods (Jiménez-Muñoz et al., 2016).

has its oldest cities located at the river margins for better accessibility, sites' shapes are usually very flat, demanding attention to soil capacity to absorb rainwater and river floods; at these cities the permeability of soil is strategic to allow vertical infiltration, especially at the portions of the site that are part of rivers' floodplains.

However, from a market perspective, in such a context, the best locations are the higher portions of the city's site, which are often those benefited from official investments that produce differential income to the extent they concentrate technical networks and social infrastructure, but also cause impermeability of soil and deforestation. The reduction of permeable soil directs all the rainwater to the available drainage systems, and even when it is accepted as technically adequate, it discards a large volume of water at once to the lower areas of the cities, that should not be occupied (Schultzer, 2014) but actually are the densest ones, prompting floods. The hegemonic conception of drainage assumes that very expensive macrodrainage interventions are solutions to these circumstances that depend on billionaire bank loans to city administration, despite the prevailing municipalities' economic indicators. This is a postoccupation solution based on highly complex technologies, which allows the incorporation of consolidated previously informal occupation to the formal market. It is a trajectory that triples the impact on the poorest groups—the first impact occurs when they are forced to migrate to the city outskirts, or to floodplains, and establish a periurban livelihood, mixed of rural and urban activities, or when the city sprawl reaches their original settlements; a second impact occurs when they are affected by water contamination and violent build and people density increases; and a third impact occurs when the area suffers gentrification due to interventions.

Despite the fact that the city is produced collectively, without cost for landowners, the access to land with infrastructure and is not prone to floods, it is otherwise restricted by market, to those who can pay for land incomes after interventions (Singer, 1982). This reinforces the occupation of floodable and risky areas by the most vulnerable social groups. When there are public investments in housing, they are also affected by market determinations, and official schemes are built at the cheapest locations, distant from the consolidated city, even though it is a territory of peasants, who manage forest products and suffer the process described by Raesbaert (2010) as deterritorialization.

The housing schemes built at distant and cheap land peripheries are offered to the poorest groups, and mostly for those who are inserted in formal economy, that will struggle over decades for the extension of services and will provide the appreciation of land placed in between these settlements and the consolidated urban fabric. The combination of private ownership control and underprovision of services, equipment, and infrastructure creates an experience of exclusion for popular housing inhabitants, and a blind spot about the original inhabitants, who had practiced land uses that respect the geomorphology and the ecological features of the site.

This dynamic of city expansion was investigated by Angel et al. (2011) and Ramos (2014) by using time series to characterize the urban spatial structures through three attributes, density, fragmentation, and occupation of the urban fabric, and by discussing how they evolve between periods. This approach assumes that urban expansion is the result of dynamic processes of conversion of rural, or periurban, land uses for urban uses, and that the dynamics of land incorporation, under the focus of the market and pressure for population growth, are associated with profit expectations on the part of those who obtain economic benefits by expanding the urban fabric of cities. For the case of Belém the analysis was based on a time series of the Normalized Difference Vegetation Index to represent vegetation cover and a temporal series of urban growth patterns. The satellite images used were Landsat (sensor TM and OLI) for the years 1984, 1999, and 2018 (Angel et al., 2011; Ramos, 2014; Rouse et al., 1973).

13.3 The case of Belém

Belém is a 404-year-old metropolis that was the main port of the Amazon region for three centuries and a typical point of colonizers control overseas. The city has benefited from the exportation of several products extracted from the forest and was considered a place of trade and civilized life amidst the forest (Meira Filho, 2015 [1976]). The city had portuary and administrative functions, it has never been a place of production, as industrial European cities were, but it was a point to control the "drainage" of ready products from forest (seasoning, timber, fruits). This trajectory was reedited many times over time and was unable to foster social division of labor inside the city, at the same time that has devalued local knowledge. Two periods are remarkably considered on understanding the city expansion, the first was the rubber boom (1850–1910); the second was the military dictatorship (1964–85), when the city growth reached levels three times higher than the country average.

The site of Belém has a long-term natural dynamics, which was modeled by climate and hydrodynamics for ages (see Fig. 13.2). It was carefully selected by an indigenous people—the Tupinambás, as the best location considering navigation and control the entrance of foreigners into the continent, and taken by Portuguese in 1616, when the city was funded [Meira Filho, 2015 (1976)]. But that features of the site have also created a challenge to the expansion of Belém, considering the need to either overcome or respect small rivers and floodplains. By the 17th century the Portuguese experience of landfill was set as reference for drainage; and by the 20th century, this perspective of urbanization was unfolded into a practice of canalization and transformation of rivers in a "natural system" to collect sewage, to control floods, to improve accessibility, and to promote social cleavage (Cardoso & Ventura Neto, 2013).

During the rubber boom, despite guidance of an alignment plan from 1893 (Muniz, 1893) to fully occupy the portion of land that belonged to the city of Belém (the First Patrimonial League of Belém—FPL, its outline is in Fig. 13.3), actual occupation had avoided the floodplains and preserved creeks and inner rivers, up to 1960s, and these areas were occupied by rural land uses (orchards for self-consumption, animal breeding, petty extractivism). Since the Brazilian military dictatorship, densification

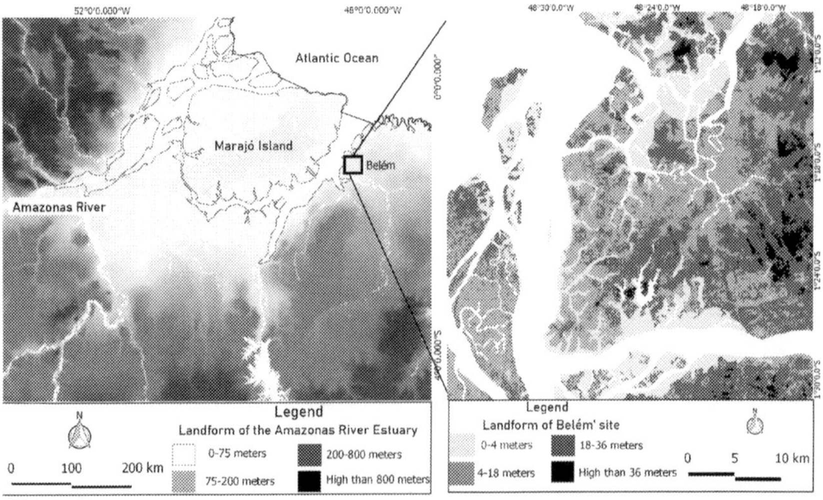

Figure 13.2 Site features (landforms) of the Amazonas River estuary, location of Belém, and landforms of Belém and its metropolitan region. *Credit: USGS, 2014; CPRM (2015). Susceptibility Cards to Gravitational Mass Movements and Floods. Belém: CPRM, 2015. [Digital cartography, shapefile format].*

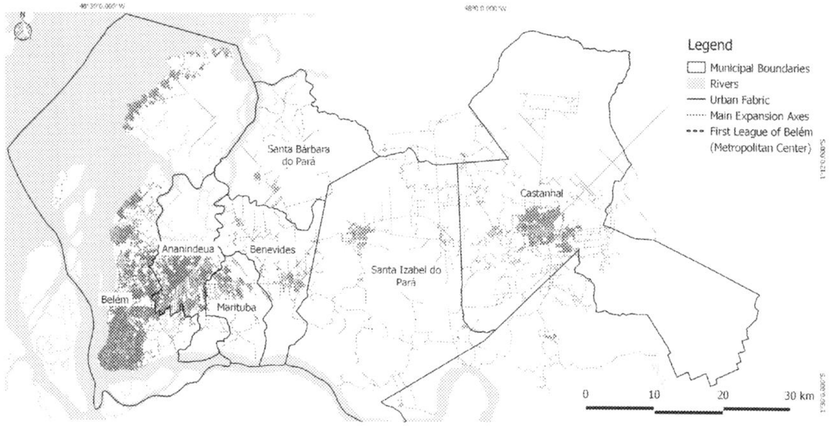

Figure 13.3 Present composition of Belém metropolitan region, with limits of municipalities, main axes of expansion, and First League of Belém (metropolitan center). *Credit: IBGE (2018). Delimitation of Brazilian cities. Rio de Janeiro: IBGE, 2018. [Digital cartography, shapefile format]; IDESP (2012). Road network in the Metropolitan Region of Belém (MRB). Belém: IDESP, 2012. [Digital cartography, shapefile format]. Elaboration: Thales Miranda.*

of the FPL is occurred by disruptive land occupation practices: occupation of middle blocks backyards by popular housing, high raised buildings for upper class housing, and incremental landfill of floodplains by excluded social groups. The former railway (built during the rubber boom to connect Belém to the Atlantic coast) was replaced by a federal road (opened in 1958 to connect Belém to Brasília), creating an axis of urban expansion and conversion of rural into urban land, that has also structured the municipalities that take part in present metropolitan region of Belém. There was an extension of the railway toward North that became the axis of Belém's expansion (see Fig. 13.3). Initially, rural migrants established in Belém had a possibility of adaptation to the urban lifestyle, practicing rural uses in the floodplains, it was a resilient solution in environmental and social terms. However, the increasing urbanization turned land into a commodity and prompted the informal densification of these areas, transforming them into ordinary peripheries.

An examination of urban regulations and plans delivered to urban areas of Belém (other municipalities did not have regulations up to the 2000s) against actual urban fabric expansion shows that planning trends found in official documents were not connected with natural features of the Belém's site. Central government policies and subsequent local government regulations, especially for urban development, have left out from their rationale, the need to properly tackle geomorphological features such as respect to water flows and their connections with the Amazon landscape. Because of that, Belém is losing the features that have made it an inclusive city over time, where different systems of access to land were possible, due to the public nature of land, to the availability of natural resources (several inner rivers, proximity to forest, soil available to domestic and publicly accessible orchards).

Official documents (plans, laws, ordinances) have a discourse that does not recognize the existing settlements and their means to live. Guidance was devised as if market-oriented built forms would replace urban fabric without respect to the site and environmental features. The emphasis of the policies has been to stimulate civil construction without providing instruments for more holistic urban policy. There was no long-term planning to provide guidance about how housing developments should be located, and this has to do with the nature of the regulations in Brazil, where regulations are nearest to the American tradition of zoning. However, the system is inefficient due to problems in the management structure alongside weaknesses in the regulations sets to control development with respect to environmental peculiarities, aggravating the gap between policy design and actual implementation.

The 1980s were remarkable times for urban regulations in Belém, when comprehensive planning, based on zoning, helped in setting up the conditions for reserving land for real estate. It was the decade that Belém implemented its own *Lei de uso do solo* (Land use law) and *Lei de parcelamento* (Land parcelling law), following the administrative model of Brazilian municipalities. These laws, after approval by the City Council, constitute the two basic urban design control instruments. The first was comparable to a zoning regulation because of the basis on which it is drawn up, although more attention was given to density regulation than to land use distribution. The final physical form of the housing units was decided upon by planners responsible for granting planning permissions according to density and land use regulation; this includes, for example, minimum dimensions and specified building materials.

Metropolitan urban planning had a functional approach to zoning and parceling, which were the basis of municipal laws, but this framework was abandoned after the 1988 Constitution. At that time, the city was organized through mass housing projects built on peripheral locations, according to a "leap-frog" pattern, that have created gaps in the urban fabric that favored real-estate agents in the following decades. During the 1970s, they were based on relocation of residents from informally occupied floodplains to social settlements as part of what continued in the 1980s, a decade characterized by actions for urban development plans upon massive infrastructure provision and land use optimization, when relocation of residents was then associated with sites and services. The construction of peripheral housing settlements has reduced vegetation cover that was reduced to the border of the most urbanized area.

After the military dictatorship, housing policies have emphasized partnerships between public and private sectors, mainly because of the failure of the National Housing Bank (BNH in Portuguese), in funding housing and infrastructure needed for upgrading already existing social housing developments. Due to insufficient and inadequate regulation on the action of the private sector, funds were used by companies to build housing developments directed to higher income classes that could pay for their mortgages. Macrodrainage works were conducted in the inner city rivers and creeks, to allow building densification in occupied floodplains and improve accessibility in the city. This action has aggravated site impermeability, and over time it was noticeable that market-oriented policies have resulted in different patterns of occupation with emphasis on built densification (infill occupation) and urban expansion as a sprawl, which have further compromised natural drainage.

The combination of official interventions and market-oriented occupation has resulted in the impermeability of the area that should have the most effective absorption of rainwater through vertical infiltration. The former generous backyards and rural land uses practiced across floodplains had an important hole to cope with the high levels of rain regimes. It is important to inform that a large amount of FPL land was public and permissions had costs when granted to occupiers in Belém until 1988 Constitution, when private property was established as the main fundiary regime. However, despite the land was public, it had been possible to claim land tenure of floodplains before this change; rich families used to do so to use land as a guarantee of bank loans during the rubber boom and were strongly benefited by the change of tenure regime and by the work done by excluded social groups on flooded areas (landfill and political bargains for landfill and macrodrainage works of these lands) (Benatti et al., 2017).

Fig. 13.4 shows how built densification of the FPL of Belém was intense over the 1980s, with intense replacement of backyards by newly

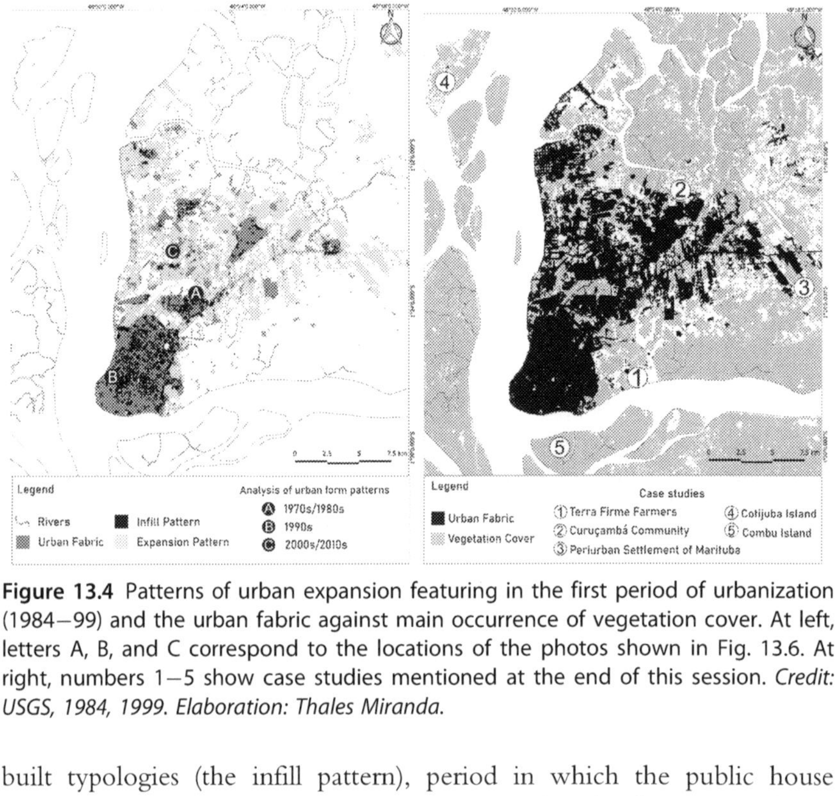

Figure 13.4 Patterns of urban expansion featuring in the first period of urbanization (1984–99) and the urban fabric against main occurrence of vegetation cover. At left, letters A, B, and C correspond to the locations of the photos shown in Fig. 13.6. At right, numbers 1–5 show case studies mentioned at the end of this session. *Credit: USGS, 1984, 1999. Elaboration: Thales Miranda.*

built typologies (the infill pattern), period in which the public house schemes had created new vectors of city sprawl and valorization of land (former leapfrog extensions, already shown as discontinuous but consolidated urban fabric), at the same time that floodplains in the city center (FPL) were occupied and are marked as extensions. During the period between 1984 and 1999, new areas for expansion were open, intensifying the continuity of urbanization, despite all sites requirements for permeability. Only institutional areas (large properties that belong to federal institutions, to the Army and Air Force) were not occupied, as well as bigger internal rivers' floodplains located at the expansion of the city of Belém.

Fig. 13.5 shows how the process of expansion was repeated in the area marked by the pattern of expansion at the previous period, due to the amalgamating of official housing schemes to other forms of land formal development and also of informal production of settlements. At this stage, long-term institutional areas were reduced, thanks to transfers of their land to private market. The two axes of expansion (Northwards and

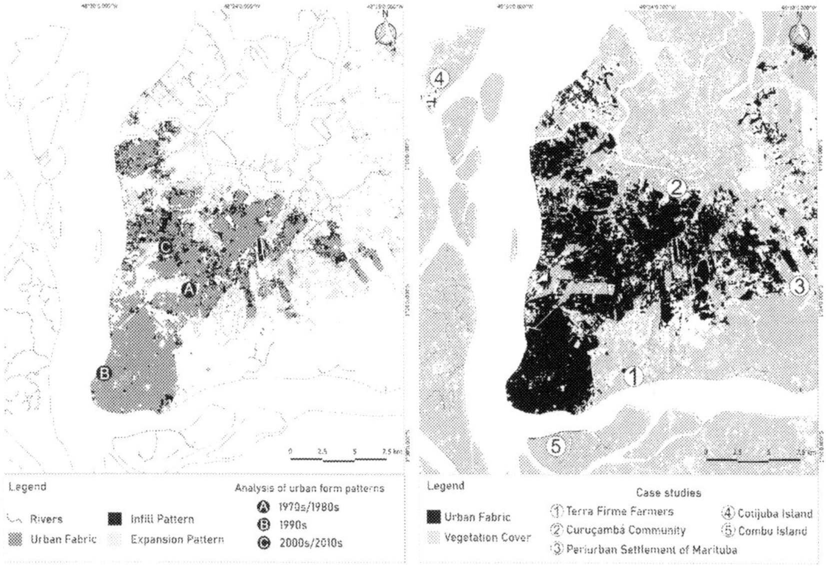

Figure 13.5 Patterns of urban expansion featuring a second period of urbanization (1999–2018). *Credit: USGS, 1999, 2018. Elaboration: Thales Miranda.*

Eastwards), which follow the two directions of a former railway turned into roads, reproduced the process of vulnerability to floods already observed at the city center but created over four centuries, within a shorter period.

At the 21st century, despite the global claim for sustainable policies and an awareness of climate change impacts (IPCC, 2014), in the city of Belém, and within its metropolitan region, there was a continuity of official practices that deliver privileges for the market, despite the need to periodically review planning and regulations and the possibility to articulate planning across the other neighboring municipalities. Persisting shortage of housing units was a direct consequence of policies inadequacy and shortage of funding to translate objectives and plans into actual houses, neighborhoods and suitable sites' land uses. Public policies still assume that all inhabitants are inserted in formal economy and do not consider the social, economic, and environmental potential of urban agriculture and other land uses capable to occupy people and to maintain soil permeability; instead of it, they keep the location of mass housing schemes on cheap land, far from the concentration of jobs, without planned accessibility by public transport to inner city areas and where jobs and services continued to be centralized.

Fig. 13.6 presents a summary (and a timeline) of the different planning instruments devised in Belém to guide urban development, assessing environmental impacts and resulting urban form patterns.

Along this speculative process of production of urban space, urban farmers and petty extractivists were gradually expelled from the flooded areas of central Belém to the metropolitan expansion area. In these new areas, water and soil are less contaminated and it was possible to establish the production of small-scale food as a family income, creating local food

Decade	Urban planning instruments	Objectives	Environmental orientation / impact	Urban form patterns
1970s and 1980s	*Planos diretores* were made at the metropolitan scale, including Zoning laws and proposals of drainage works (CODEM, 1975; PMB, 1988, 1988a, 1988b)	Comprehensive planning based on zoning devises land development and infrastructure for land use optimisation	Green areas reduction due to implementation of peripheral social housing esquemes, and informal occupation adjacent to official housing. Contamination of rivers by sewage. Macro drainage and building densification in the floodplains	(A) Pattern of urban expansion: public peripheral social housing and informal occupation. 
1990s	After the 1988 Constitution planos diretores became municipals and complemented by zoning and land occupation laws (PMB, 1993, 1999)	Constitutional determinations: social function of the city for urban reform (target: to reduce real estate speculation)	High rise buildings causing water proofing land at the centre. Reduction of backyards,occupation of floodplains (look behind verticalized area). Vegetation cover reduced to protected areas.	(B) Pattern of infill: verticalization at the city center, infill urban expansion pattern. 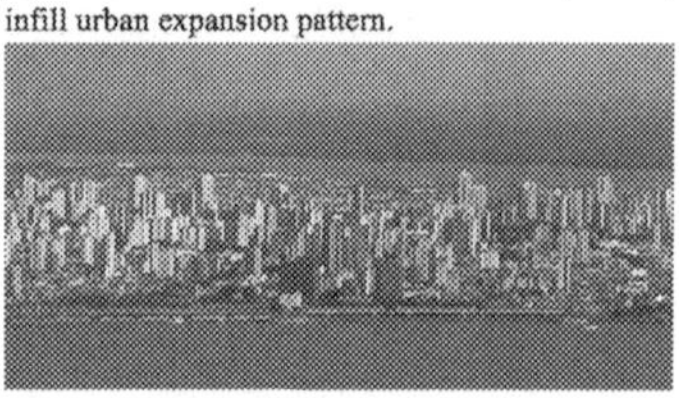
2000s	City statute (Brasil, 2001)	Neoliberal trends in zoning laws threatened social function instruments	Urban sprawl and consumption of waterlogged areas, water contamination, lack of control of vegetation cover loss.	(C) Private closed vertical housing condominium, forming urban expansion and infill of former urban expansion patterns
2010s	*Plano Diretor* revision, new generation of social housing programs (PMB, 2008).	Real estate deregulation, financiarisation of housing schemes within housing programs	Verticalization of former urban expansion patterns, following accessibility potentials. Progressive occupation of non protected land still covered by vegetation.	

Figure 13.6 Urban planning instruments—objectives and outcomes. *Credit: Sources: Image A: COHAB, PARÁ, 1971; Image B: Lima, 2019 and Image C: Mota, 2018.*

sovereignty, promoting local gastronomic identity, and providing a balanced relationship with natural resources (Cardoso et al., 2019). Cases studied shown in Fig. 13.7 present the spatial arrangement of five different

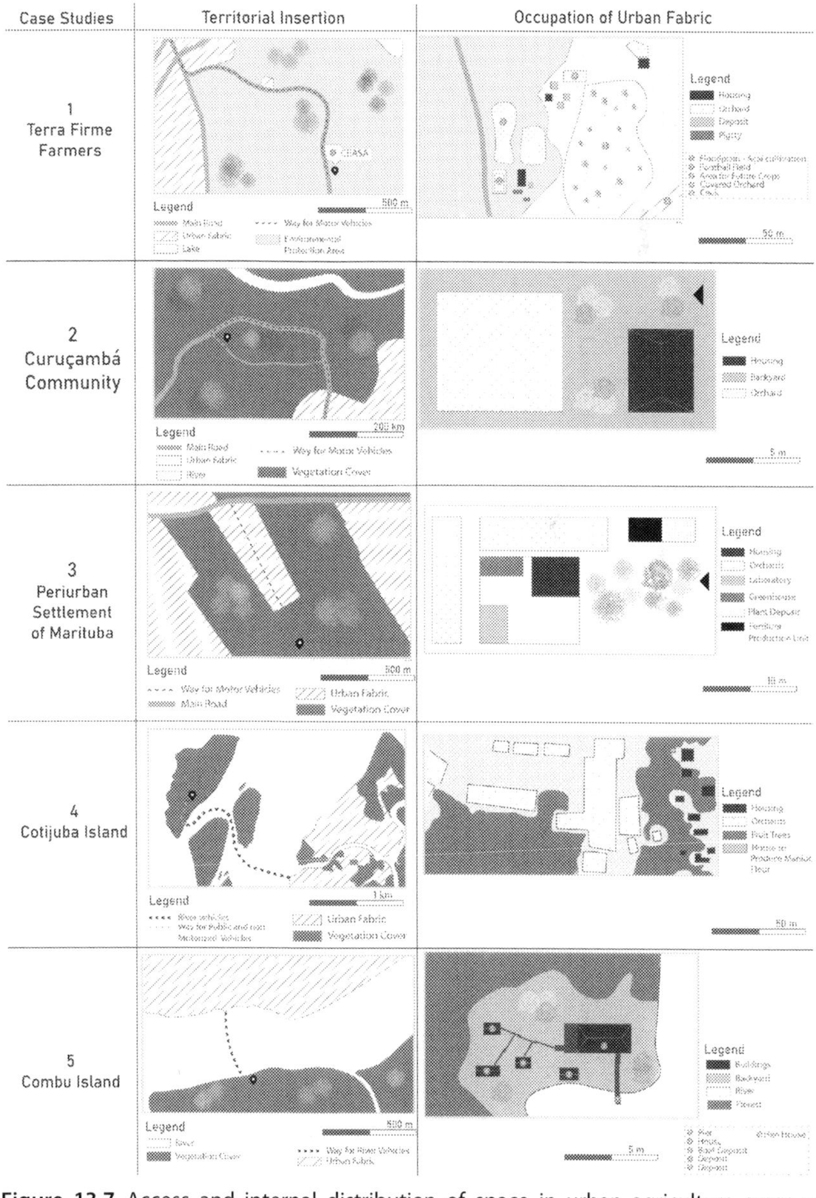

Figure 13.7 Access and internal distribution of space in urban agriculture areas of Metropolitan Area of Belém. *Credit: Elaboration: Elis Brazil.*

cases, to illustrate the different contributions from organization of contemporary urban farmers in Belém. They are placed in the continent at the borders of Belém urban fabric and also in its islands. The case study 1 recovers a resettlement of floodplains' producers (indicated in Figs. 13.4 and 13.5), which were relocated next to the Central Estadual de Abastecimento—CEASA (Pará State Central for Food Supply)—and were forbidden by Belém City Hall, to live next to the production area and treated as if they were retail and services workers. It indicates the official misunderstanding of these urban farmers' daily arrangement that demanded to live and to produce at the same site. The only remaining family of this experience only succeeds after settling in the site, the other gave up due to burglary and excessive costs with mobility.

Case number 2 (also indicated in Figs. 13.4 and 13.5) is located at Curuçambá. It features the settlement that has received part of producers from the floodplains and other migrants. They manage to provide organic vegetables to open markets in Ananindeua and part of Belém, benefiting from a wide diversity of socio-spatial practices. These workers moved to the edge of Belém and Ananindeua municipalities (Cardoso & Miranda, 2018), being pushed to land located beyond official housing schemes, where there was clean water and land available, despite the low confidence of young generations in the future of such activity.

Number 3 (indicated in Figs. 13.4 and 13.5) refers to a small farm in Marituba, a municipality that has the largest number of urban farmers (actually periurban peasants) settled within the metropolitan region, but does not acknowledge rural areas in its territory favoring real-estate sector demands. At the edge of a former land invasion that had half of its area dedicated to housing and half to *sítios* (small farms), there is a very successful case or urban farming. The size of the plot is equivalent to the ancient plots located in the formal areas of FPL, with generous backyards. The case studied is connected to other units of production (other *sítios*) and to the association Pará Orgânico, which provides organic products to the two most important open markets of organic food of Belém.

Numbers 4 and 5 correspond to the community of Cotijuba Island, where local farmers supply schools and the open market, and to Combu Island, which has a very prone to floods site, where producers were organized to supply cocoa to exclusive restaurant of Belém. The islands are formally considered rural settlements for extractivism and familiar production, but it has suffered from strong pressure to be occupied by settlers who brought other land uses due to the proximity to the continent.

13.4 Final remarks

According to the State Institution for Rural Assistance (EMATER), there were 2379 rural producers formally listed in 2019 (PROATER, 2019), within the metropolitan region of Belém. These are the cases formally acknowledged, assisted by EMATER; however, there are more cases of urban agriculture activities outside that number. According to the participants of the movement Pará Orgânico, which includes 40 families and is the most organized association or urban farmers in the metropolitan region, they do not have official assistance, apart from the control of practices to be considered organic agriculture by local government. In Combu Island, for instance, the production of chocolate from terroir (native) varieties of cocoa and açai palm tree has created a return of inhabitants' children who were working in the continent, to help in family business connected to local gastronomy.

Data on these new formats of endogenous-based practices are underestimated. They are understood as ecologically responsible, to the extent, they reedit the ancient paradigm of biodiversity management and of systemic connection between rivers, vegetation, soil, biodiversity, and people, in either available or prone to floods land. And despite the pressure of hegemonic sectors to maintain predatory ongoing urbanization, plus the lack of official policies' commitment to foster adaptation of cities to a context of climate change and to the need of reduction of emissions in Brazil, these practices offer a window to an underlying local capacity to move toward these aims, thanks to everyday resistance of poor inhabitants to give up of local identity and knowledge.

Besides that, in 2015 Belém has received the title of one of the World Cultural Capitals of Gastronomy, named by UNESCO, and most of the support for that comes from the silent and persistent work of local urban farmers. However, there is a conspicuous competition between the different consumers of land and between different ways of lives, with the loss of natural resources as the main result that jeopardizes this sociobiodiversity. Under this competition, it is harder to provide popular housing and nonlucrative land uses of communitarian interest, as orchards, to promote social inclusion and to increase soil permeability.

The successive cycles of consolidation within the metropolitan area tend to decrease in quality in each new turn. Beyond that, to sustain "rural" uses inside, the city is not even acknowledged as a possibility by public policies, and the more traditional livelihoods that depend on soil, biodiversity, and clean water continue to be expelled to outer rings, opening space for an intensification of urbanization that increases environmental vulnerability in a time in which their permanence should be encouraged, following global search for strategies to increase resilience for all social groups.

In this sense the disentanglement of layers of formal, informal, and endogenous land occupation was explored in this chapter to better expose Belém's potential to the provide basis for an alternative relationship between humans and nature, to hegemonic practices based in industrial Western societies. To promote this, epistemological turn can be an effective local solution to postmodern demand of respecting diversity and next-generation right for cities but also for a planet. It also allows a reflection on the importance to say to whom is addressed an action or a paradigm, of either sustainability of resilience, to tackle environmental crises.

References

Angel, S., Parent, J., Civco, D., & Blei, A. (2011). *Making room for a planet of cities. Lincoln Institute of Land Policy, policy focus report.* Brasil, Federal Law no. 10257/2001. Estatuto da Cidade (in Portuguese).

Benatti, J., Fischer, L. (Coord), Rosso, C., Silva, E., Santos, L., & Brigida, N. (2017). *Análises bibliográficas, legislativas e jurisprudência para apuração do remanescente patrimonial.* São Paulo: Acquerello (in Portuguese).

Becker, Bertha (2013). *A urbe amazônida.* Rio de Janeiro: Garamond.

Cannigia, G. F., & Maffei, G. L. (1979). *Lettura dell'edilizia de base.* Veneza: Marsilio Editori. (in Italian).

Cano, W. (2011). *Ensaios sobre a crise urbana no Brasil* (pp. 181–205). Campinas: Ed Unicamp. (in Portuguese).

Cardoso, A. C. D., Vicente, L. R., & Oliveira, R. C. B. (2019). Os invisíveis portadores de Futuro. *Cadernos de Pós-Graduação em Arquitetura e Urbanismo, 19*(1), 23–37. Available from https://doi.org/10.5935/cadernosarquitetura. (in Portuguese).

COHAB/PARÁ. Companhia de Habitação do Estado do Pará. (1971). *Photograph of Cidade Nova Settlement in 1971.* Not published.

Cardoso, A. C., & Miranda, T. (2018). Invisibilidade social e produção do espaço subordinado em Belém (PA). *Paisagem e Ambiente, 41*, 85–107. Available from https://doi.org/10.11606/issn.2359-5361. (in Portuguese).

Cardoso, A. C., & Ventura Neto, R. S. (2013). A evolução urbana de Belém: trajetória de ambiguidades e conflitos socioambientais. *Cadernos Metrópole, 15*(29), 55–75. (in Portuguese).

Costa, F. (2019). *A brief economic history of Amazon (1720–1970).* London: Cambridge Scholars Publishing.

Farret, R. L. (1985). Paradigmas de Estruturação do Espaço Residencial Intra-Urbano. In R. Farret, et al. (Eds.), *O Espaço da Cidade: contribuição à análise urbana* (pp. 73–90). São Paulo: Projeto. (in Portuguese).

Ferreira, J. S. W. (2007). *São Paulo: o Mito da Cidade-Global.* Rio de Janeiro: Vozes. (in Portuguese).

Forman, R. (2014). *Urban ecology, science of cities*. London: Cambridge University Press.
Gonzales, S. F. N. (1985). A Renda do Solo Urbano: hipóteses de explicação de seu papel na evolução das cidades. In R. Farret, et al. (Eds.), *O Espaço da Cidade: contribuição à análise urbana* (pp. 73–90). São Paulo: Projeto. (in Portuguese).
Hagan, S. (2015). *Ecological urbanism, the nature of the city*. London: Routledge.
Heckenberger, M. J., et al. (2008). Pre-Columbian urbanism, anthropogenic landscapes, and the future of the Amazon. *Science (New York, NY), 321*(5893), 1214–1217. Available from https://doi.org/10.1126/science.1159769.
IPCC. (2014). *Fifth assessment report: Impacts, adaptation and vulnerability*. New York: Cambridge University Press.
Jiménez-Muñoz, J., Mattar, C., Barichivich, J., et al. (2016). Record-breaking warming and extreme drought in the Amazon rainforest during the course of El Niño 2015–2016. *Scientific Reports, 6*(33130). Available from https://doi.org/10.1038/srep33130.
Lefebvre, H. (1991). *The production of space*. Oxford: Blackwell.
Maricato, E. (2000). As ideias fora do lugar e o lugar fora das ideias. In O. Arantes, et al. (Eds.), *A Cidade do Pensamento Único: desmanchando consensos* (pp. 121–192). Petrópolis: Vozes. (in Portuguese).
McHarg, Ian (1969). *Design with nature*. New York: Doubleday/National History Press.
Meerow, S., Newell, J. P., & Stults, M. (2016). Defining urban resilience: A review. *Landscape and Urban Planning, 147*, 38–49. Available from https://doi.org/10.1016/j.landurbplan.2015.11.011.
Meira Filho, A. (2015 [1976]). *Evolução Histórica de Belém do Grão-Pará. Fundação e História 1616–1823*. Belém: M2P (in Portuguese).
Mota, J. M. (2018). *Mudanças no processo de produção do espaço belenense diante da financeirização* (Dissertation). Belém: UFPA (in Portuguese).
Muniz, J (1893). *Patrimônios dos Conselhos Municipaes do Estado do Pará*. Lisboa: Aillaud & Cia.
Neves, E. G. (2006). *Arqueologia da Amazonia*. Rio de Janeiro: Zahar. (in Portuguese).
PROATER. (2019). *Programa de Assistência Técnica e Extensão Rural*. Belém: EMATER (in Portuguese).
Raesbaert, R. (2010). *Regional-global, dilemas da região e da regionalização na geografia contemporânea*. Rio de Janeiro: Bertrand Brasil. (in Portuguese).
Ramos, F. R. (2014). *Três Ensaios sobre a Estrutura Espacial Urbana em Cidades do Brasil Contemporâneo: economia urbana e geoinformação na construção de novos olhares* (Thesis). São Paulo: FVG (in Portuguese).
Rees, W. (1992). Ecological footprints and appropriated carrying capacity: What urban economics leaves out. *Environment and Urbanization, 4*(2), 121–130. Available from https://doi.org/10.1177/095624789200400212.
Rolnik, R. (2018). *Guerra de Lugares*. São Paulo: Boitempo. (in Portuguese).
Rouse, J. W., et al. (1973). Monitoring vegetation systems in the great plains with ERTS. In: *Earth resources technology satellite-1 symposium* (pp. 309–317). Washington, DC: NASA.
Schultzer, J. G. (2014). *Cidade e Meio Ambiente, a apropriação do relevo no desenho ambiental urbano*. São Paulo: Edusp. (in Portuguese).
Singer, P. (1982). O uso do solo na economia capitalista. In E. Maricato (Ed.), *A produção capitalista da casa (e da cidade) no Brasil industrial* (pp. 21–36). São Paulo: Alfa Omega. (in Portuguese).
Swann, W., & Deslatte, A. (2019). What do we know about urban sustainability? A research synthesis and nonparametric assessment. *Urban Studies, 56*(9), 1729–1747. Available from https://doi.org/10.1177/0042098018779713.

USGS. (1984). *Landsat 5 (TM) image*. Washington, DC: USGS. [Satellite image]. <https://earthexplorer.usgs.gov> Accessed September 2019.
USGS. (1999). *Landsat 5 (TM) image*. Washington, DC: USGS. [Satellite image]. <https://earthexplorer.usgs.gov> Accessed September 2019.
USGS. (2014). *Shuttle Radar Topography Mission (SRTM) image*. Washington, DC: USGS. [Radar image]. Available from: <https://earthexplorer.usgs.gov> Accessed September 2019.
USGS. (2018). *Landsat 8 (OLI) image*. Washington, DC: USGS. [Satellite image]. <https://earthexplorer.usgs.gov> Accessed September 2019.

Further reading

CODEM – Companhia de Desenvolvimento e Administração da Área Metropolitana de Belém, BNH – Banco Nacional da Habitação and DS - Desenvolvimento e Sistemas. (1975). *Plano de Desenvolvimento da Grande Belém PDGB*. Belém: CODEM/BNH/DS (in Portuguese).
McHarg, I. (2000). *Proyectar con la naturaleza*. Barcelona: Editorial Gustavo Gilli. (in Spanish).
PMB – Prefeitura Municipal de Belém. (1988a). Law n. 7,399 11/01/88. Dispõe sobre o Parcelamento do Solo do Município. *Diário Oficial do Município Belém: Imprensa Oficial, 19*(6), 233. (in Portuguese).
PMB – Prefeitura Municipal de Belém. (1988b). *Law n. 7,400 25/01/88*. Belém: Imprensa Oficial (in Portuguese).
PMB – Prefeitura Municipal de Belém. (1988c). *Law n. 7,401 29/01/88*. Belém: Imprensa Oficial (in Portuguese).
PMB – Prefeitura Municipal de Belém. (1989). *Law n. 7,452 04/07/89*. Belém: Imprensa Oficial (in Portuguese).
PMB – Prefeitura Municipal de Belém. (1993). *Law n. 7.603/93. Plano diretor de Belém*. Belém: Imprensa Oficial (in Portuguese).
PMB – Prefeitura Municipal de Belém. (1999). *Law n. 1*. Belém: Imprensa Oficial (in Portuguese).
PMB – Prefeitura Municipal de Belém. (2008). *Law n. 8.655 30/07/2008 Plano diretor de Belém*. Belém: Imprensa Oficial (in Portuguese).

PART IV

Perspectives in boosting practices' contribution towards sustainability, adaptation and resilience

CHAPTER 14

Toolboxes for improving the stabilization of the resilience practices

Giulia Pesaro
REsilienceLAB/Co.O.Pe.Ra.Te. Ldt, Italy

14.1 Enhance the diffusion and duration of resilience practices. Tools typologies for enhancing the achievement of resilience goals

The issue of tools to support the transition of territorial systems and local communities in terms of resilience and sustainability is delicate and complex. The main reason is to be found in the fact that the resilience of communities and territories potentially involves all sectors and all intervention fields in relation to both the activities that normally take place in a territorial system and the needs that emerge over time in the face of changes and developments determined by endogenous and exogenous dynamics, shocks, and crises.

The tools here referred to are those through which it is possible to make feasible and implement the local level practices and, at the system level, the policies and the governance choices aimed at strengthening resilience recognized as a medium- and long-term operational objective. These are therefore functional tools activated for achieving goals which have been already identified and defined in the previous chapters as coherent and consistent with the concept of operational territorial and community resilience. This is an overall framework of governance goals and setting of strategic and territorial plans and programs for the production and enhancement of territorial and community resilience.

Three main discussion elements should be carefully considered in discussing the topic of the "resilience action tools" in a scale-up perspective: the goals, the actors, and the need for stabilization.

[ECO]systems of Resilience Practices
DOI: https://doi.org/10.1016/B978-0-12-819198-9.00017-X

14.1.1 Tools and resilience goals

A first element of the reasoning relates to the goals, which in turn are related to a specific concept or definition of resilience in terms of the "territorial and community quality to be achieved." Resilience, in this volume, is in fact an intrinsic characteristic of the system that we intend to achieve to strengthen the stability and the development level of the systems themselves over time. A concept of resilience, by also dealing with physical and immaterial and quantitative and qualitative territorial resources and endowments, incorporates the concept of sustainable development.

The link with local specificities is very strong, as well as the involvement of associations, administrations, and institutional bodies in well-defined physical and environmental spaces.

The same happens when looking at the target—action fields. The distinction in different action fields, such as food systems, urban environments, urban commons, and ecosystems, is useful because it allows to study in detail the related problems and find tailored solutions that are able to maximize the action feasibility and effectiveness. On the other hand, practices developed in a multidisciplinary perspective, dealing with a variety of action fields, might be disadvantaged, even if they better fit goals of systemic resilience. For instance, education, training, and communication actions should be activated in a more stable and scale-up perspective. Not only a focus activity for the feasibility of individual practices but also an opportunity for enhancing the awareness about the values of territorial and community resilience, and the exchanges and collaboration leading to creativity, innovation and codesign, and coproduction.

14.1.2 Tools and resilience actors

A second reasoning element is the complexity linked to the typologies of subjects who use or offer intervention tools. One of the elements that emerged from the reading of the many practices analyzed in the previous chapters shows a so wide variety of active subjects that it cannot be easily synthesized and reorganized into categories of users without losing the specificities and links that characterize the different territorial realities: for instance, administrative and governance organizational features and the features and dynamics of local communities. A variety, on the one hand, is one of the strengths of the concrete feasibility of the practices but, on the other hand, makes the action more complex at the system and over local levels.

Variety is one of the strengths of operational resilience and is one of the elements which seems to characterize the concrete feasibility of the practices at each reference territorial level. The capability to promote the involvement of the whole of the local actors at different territorial levels, for instance, revealed to be an element for the success of the initiatives. On the one hand, local-based subjects are familiar with and are able to integrate that set of very specific characteristics and conditions that enable each category of targets to act more effectively and efficiently in its own action field and territorial level. On the other hand, act at the over local system level and activate the related resources make it possible to abstract fundamental elements to enrich or reframe the goals to be achieved.

A resilience is, therefore, linked also to capacity building, to the development of innovative solutions, and up to the support and promotion of codesign and coproduction practices as mentioned earlier.

It is necessary to find a balance between the tools suitable for the capillarity which is needed for strengthening the operational feasibility, implementation and stabilization of individual initiatives, and the incentives for the dissemination of system-level strategic plans, programs, and projects, generally accompanied by financing activities.

14.1.3 Tools and resilience stabilization

Resilience is almost always seen by promoters and operators as a characteristic that should be stabilized as a territorial and community quality. On the other hand, the initiatives that arise in relation to both local pressures or specific programs and strategies activated and financed by institutional bodies do not always give birth to projects capable of lasting over time or to continue to produce positive effects at the system level in the long run.

A central element of the reasoning becomes therefore that of strengthening the stability of practices over time and favoring strategic long-term programs and projects, to be developed in a sequence of coordinated and coherent implementation steps, or projects able to foster practices producing outputs and results over time, also after the end of a specific funding activity.

For this reason the issue of tools, even if challenging as already mentioned previously, needs further reasoning and resources. A couple of main questions should be considered. While developing specific activities and tools, how to strengthen the technical and scientific contents which are key resources (human and social ones as well) to enable and enlarge the capacities of the territories to better face present and future crises. Which overall strategies and toolboxes might be offered to enhance communities in the design of stable transition paths, therefore integrating resilience issues into territorial and community sustainable development actions in a more effective way (as the European Green Deal and the UN Agenda 2030 with the related SDGs suggest)?

Finally, it is important to underline that the aim of this discussion about the tools is not that of proposing a detailed classification but rather to offer a few tool categories at the system level. Resilience tools classification is a well-known and developed research field. There are many sources of literature that already exist in this sense, so many that it is not possible to cite them without risking leaving behind important contributions. Furthermore, there are literally countless projects and experiments that take place worldwide by way of countless of tools developed at the local level. For sure a strong and positive factor for resilience achievement, on the other hand, makes the identification and classification of resilience toolboxes a challenge.

The choice in this contribution is therefore to attempt identifying some basic concepts and reference macro-objectives based on which to develop operational toolboxes which best suit to overall goals of resilience practices, might them be in local-based activities or in over local support and dissemination plans and programs. A special focus will be on their contribution to stabilization and scale-up of strategic planning, programs, and practices and to the enhancement of the coherence of territorial and community actions in the context of wide-ranging actions.

In this respect the main target subjects of the following brief discussion elements could be generally identified as the "practices activators and funding institutions" which often act at an aggregate level and promote, in most cases, plans and action programs encouraging territorial actors to submit ideas and proposals coherent to the overall system of resilience goals.

Finally, the categories which follow are not intended to be exhaustive. The aim is to highlight tools that, in the light of the analyses proposed in the previous chapters, appear to emerge, in a direct or indirect way, as keys to success.

14.2 Instruction, education, and training tools

These tools aim at building the human and social resources needed for achieving resilience goals. They can be regarded as multifaced tools, as they cover different needs. First, they represent the engine of the demand of local stakeholders for the resilience enhancement, as education contributes to raising the awareness itself that resilience actions are needed for a better future of communities and territories. Second, training activities can activate and enhance the shift from intention to action, contributing to enlarging, enriching, and renovating the human and social resources and the capacity building for project design and implementation. Finally, education and training can integrate the toolbox of the individual practices as a way to enable the involved target groups to become active part of the activities and activate future initiatives.

These tools, in most of the cases, are mixed with other elements in the design of the resilience practices and are well considered as instruments able to stabilize both the individual practices and the culture of resilience, thus enhancing the transition as an overall reference scope for the whole "territorial and community resilience-oriented action."

This tool category is for sure well acknowledged but it is important to underline that, in a transition perspective, a long-term education and training strategy for making the resilience goals and practices more ordinary in territorial and community base plans is still needed.

Looking at capacity building, as Milliken says speaking about territorial resilience and peacekeeping (Milliken, 2013), another crucial element is the capability to identify and take advantage of the whole of the resources of the local systems and activate "local learning." Even if the general framework concept for resilience offered by the author is different, as he considers the activities of the peacebuilding agencies, the approach fits with the territorial resilience concept proposed in this book: "Presuming the desirability of the local resilience in question, it should be asked early on how we can encourage effective local, national and regional learning of resilience mechanisms and informal and formal means of their promotion."

Finally, it is fundamental to mention the model of "community of practices," developed by Jean Lave, anthropologist, and Etienne Wenger, computer scientist, in 1991 (Lave & Wenger, 1991) based on the introduction of the concept of learning as a social process. Wenger completed the work in 1998, which focused on the community of practices issue. The community of practices theory has been developed as the main output of a research for better understanding the learning processes of people in a "socially situated" learning framework, that is how people learned new knowledge and skills without being part of a formal training process but participating in situations where relationships, mutual understanding, and culture sharing have a special relevance (Aubrey & Riley, 2019). Participants therefore become active part of a learning community that actively takes part in teaching and transmission activities that take place around a certain topic.

It is quite easy to understand why this training and learning model seems effective for being applied to the resilience practices domain. The different social groups within a community involved in resilience practices share a common concern and may collaborate not only in fulfilling the already identified group goals but also in building the reference learning models to enlarge the local capabilities, skills, and expectations about future development. People belonging to groups, communities, of resilience practices share of course the involvement in the resilience matters. Moreover, the participants, addressed by facilitators and experts, learn from each other and might take advantage in the use of new teaching/learning technologies and tools, like, for instance, the sharing of virtual learning spaces or even games based on resilience topics.

This model will be cited also in the following paragraphs, when communication and knowledge sharing issues will be tackled.

14.3 Economic and financial-based tools

This tool category is very huge and dynamic. It goes from the most known funding programs developed to encourage the design and implementation of practices, to the very specific economic tools included in the practices and needed in terms of feasibility and implementation. Economic tools of course are important to support the activities required to obtain the expected results and to stabilize, when possible, the obtained results, taking advantage of them in the long run and transferring the models in other territorial areas according to working conditions and available resources.

If transition issues are at stake, the tools in this category would require more attention, the focus being on innovative production and consumption models, able to couple resilience goals with environmental sustainability, the diffusion of green and circular economy practices, and the design of effective tools for better facing climate change causes and impacts.

As already seen in the previous chapters (see, in particular, Chapters 2, 3, and 8), statements of maximization of efficiency and effectiveness of the resilience enhancement action appear to be crucial in stabilizing both the practices and the related results. Moreover, the economic reasoning in the project design of the activities is one of the pillars of the feasibility of the practices and of overall financing programs.

The economic-based initiatives, in the light of public- and community-based operational goals and commitments, are very often initiated by private institutions and public bodies that look at resilience as a desirable attribute of a territory and a community. Resilience could therefore be regarded as a territorial characteristic suitable to be obtained, and financed, through the recognition of the characterization of "public interest goal" and accordingly fostered and acknowledged.

In the functioning of the economic tools for supporting resilience practices and actions, the main difference from other interest fields, like the protection of the natural environments, the enhancement of sustainable development, and the action to face climate change, is the nonexistence (at least at the present time) of specific regulations, obligations, or constraints which have to be considered and reached. In general terms, here the focus is enhancing and enriching the action range, the variety, and the quantity of practices which can be activated over time in different places on a voluntary basis. As mentioned previously, territorial and community resilience action programs need to be stabilized over time to maximize the benefits coming from the use of the local resources. Economic-based tools could contribute to sustaining the activities and trigger a dynamic cumulation of the results of practices and projects over time. In this sense, economic-based instruments like financing programs performed by public and private bodies are for sure crucial and needed. Over time, the attention to inclusive and sustainable market-based instruments and over local production–consumption models might also become very important, involving a variety of territorial subjects and their interests to improve resilience policies in a solid and ordinary way, possibly integrated into a renewed relationship between demand and supply.

The action on territorial actors thus takes two possible directions: the one hand, the development of proactive economic-based behaviors and the enhancement of an action culture which looks at territorial and community resilience as an essential quality for the better functioning of a territorial system over time; on the other hand, local economic systems might acquire new tools and capabilities to better face future endogenous and exogenous shocks and crises based on the valorization of local resources. A double-sided approach, over time, might be complemented by public actions to make passive or "fatalist" behaviors expensive for private actors, like it is being done to foster action for mitigation and adaptation to climate change.

The examples of financing tools used for fostering action in programs and strategies at the institutional level (like for the Fondazione Cariplo experience and the related Observatory for Resilience Practices examined in Chapters 4, 5, and 8) as well as the very detailed economic tools which are part of the development of the different analyzed practices and case studies (like in the food sector) show that there is room for the better integration of combinations of finance-, market-, economic-, and cooperation-based tools.

Resilience goals call for the widening of the economic tools and an increase of the capabilities of the local communities to deal with them. Even if resilience action initiatives and practices have been initiated years ago, still the economic feasibility and stability of practices have been too often underestimated during the project design phases. The role of a strategic economic approach (both in financing overall programs and inside individual practices) still needs to be enhanced, to influence the behaviors of all territorial components and to steer decision-making processes by boosting the capabilities to analyze the economic dynamics and the way in which the expected results could positively influence the local economic systems in a durable way.

14.4 Monitoring, measurement, and evaluation tools

This is a very sensitive point and the discussion is ongoing (see, among others, the works of the online international workshop "Technical, social, political: an inventory of indicators of territorial resilience," held on February 6–7, 2020, and conference "Assessing Territorial Resilience: Indicators and Tools for Governance" held on June 15–16, 2020). On the one hand, the evidence of the capability of the programs and projects to reach the expected results in terms of resilience enhancement is a need, on the other hand, the debate about the measurement and evaluation tools is very heated.

Monitoring systems applied to individual practices or well-identified action programs are fundamental tools for observing the capability of the taken commitments and the involved resources in achieving the expected results. A well designed and implemented monitoring system, integrating direct and indirect performance indicators, contributes to comparing the dimensions of the inputs, the efforts, and the implementation capabilities with the outcomes and outputs, thus enabling a better understanding of the development phases and the learning from the developed models of action and experiences.

When the measurement goals directly refer to the overall enhancement of the resilience characterizing a certain territorial area and the related community, large or small ones, the complexity explodes. The measurement of the capability of a system to reach or enhance its own resilience can be regarded as a challenge which starts in the project design phase because of the great diversity in the understanding of the meaning itself of territorial and community resilience and of the even greater number and diversity of the related strategies and practices. Monitoring and evaluation indicators need to be identified according to the practices' expected results and measurement approaches and tools' setup. Moreover, resilience monitoring and evaluation tools are going to be difficult to develop because system complexity makes it challenging for observers to attribute causality and intentionality (Milliken, 2013).

An effective assessment tool should comprehensively address multiple dimensions of community resilience, take into account the connections between different spatial scales, be able to measure changes across temporal scales, develop suitable measures for capturing uncertainties, be developed and implemented in collaboration with stakeholders, and lead to development of action plans for enhancing resilience (Sharifi, 2016). Furthermore, it is difficult to completely identify, understand, and measure territorial resilience without considering the influence of geographic, territorial, social, and temporal scales. Finally, cross—scale relationships are largely neglected and communities are often being assessed as stand-alone and isolated entities (Sharifi, 2016).

Another element for the selection of assessment tools is the awareness about the need to better acknowledge the fact that resilience building is a dynamic process. In addition to relying on historical trends and baseline conditions, it is essential to monitor conditions at regular intervals and also use modeling and projection techniques to avoid being overwhelmed by constantly changing of the reference territorial framework and community (Sharifi, 2016).

Finally, another challenge would be to develop assessment tools that are flexible enough to be applicable to various locations and transferred among different practices. A good balance should therefore be achieved between the need for tailoring the action on the local specificities and the need for comparing different experiences over time and in space.

Despite the highlighted criticalities and complexities, the need to enable activators and implementers to understand the benefits obtained from the set of the made investments. Territorial systems and the related communities should have available reasoning elements and tools to acknowledge the effectiveness and efficiency of the investments of territorial resources toward resilience enhancement.

14.5 Tools for the enhancement of direct participation, communication, and diffusion of information on resilience action and practices

The assertion that the enhancement of participation, communication, and diffusion of information and culture about territorial and community resilience is a crucial factor should, at this point, be a well-established point. In response to the environmental and social challenges of an uncertain future, the action toward resilience building and enhancement can be regarded as a stabilized topic. However, there is evidence of the increasing need not only for the enhancement of the exchanges in experiences and action programs but also in the building of opportunities and spaces where to share, exchange, and increase the impact of the existing practices and foster the transferring and adaptation of the success strategies.

In the last decade the role of digital tools and media and the virtual spaces opened in the net revealed to be very effective for scaling up local resilience practices and enable connections and knowledge sharing across locations, communities of practices, action programs, and plans.

It would of course be too long, and perhaps not useful, to analyze in deep the functioning and the effectiveness of such tools in the resilience realm, as well as to underline the problem of the quality and reliability of the contents, because these factors are transversal to every kind of issue.

On the other hand, it is worthwhile to focus attention on tools which might foster capacity building and sharing to produce wider and stronger [eco]systems of resilience practices, continuously evolving and diffusing across communities and policy actions at different territorial levels.

The detail and the locally based initiatives have therefore to enlarge their boundaries and become part of a common over local repository of experiences to be considered a resource to encourage and address new and innovative future actions.

Community of practices, already mentioned earlier, can represent an effective tool in this regard, as members of a community of practice are practitioners (Wegener, 1998). They develop a shared repository of resources, available by way of different technologies and incorporated in different "knowledge objects": narratives, articles, reports, images, practices outputs, virtual spaces associated with the practices.

Scientific research and analysis is increasingly involved and evolving as a system of resources available at the over local level and able to "extract" lessons learnt in an effective and efficient way. Still, to make research and technical-based analysis available to a larger audience, some more interventions are needed. Enabling knowledge, communication, and participation in the territorial and community resilience domain requires a strong joint effort of scientists, practitioners, decisions makers, educational and civil society stakeholders with the common goal to develop appropriate skill sets and solutions.

A second assumption is the concept of knowledge building, transmission, and coproduction as a multidisciplinary and multiscale space of interaction—a space where subjects producing, diffusing, and using knowledge interact with each other exchanging knowledge resources in a real or virtual space according to cyclic chains of production–supply–demand for knowledge, like in a marketplace enabling coproduction and codesign of practices and other knowledge sharing and transmission tools and platforms (Mejri & Pesaro, 2015). All typologies of actors continuously interact in a dynamic environment, exchanging and matching knowledge embedded in different available tools while contributing in the development of new operational solutions and practices or in the transfer and adaptation for different users. Proximity to end users is also an important element, as the maximum effect in enhancement, diffusion, sharing, and enabling of the resilience culture is near to the field, at the territorial and community operational levels. Analysis and complex formal modeling, which characterize academia and research, is of course fundamental in improving the understanding of the resilience matters, the modeling of resilience feasibility conditions, and the monitoring and measurement activities but complexity has to be solved in approaches and methodologies when addressing practices involving territorial stakeholders (Mejri & Pesaro, 2015).

This reasoning opens up spaces for in-depth studies to increase the effectiveness and efficiency of collaborative tools such as codesign and coproduction of resilience practices, as well as to citizen science. As far as resilience is a territorial quality which can be addressed by mobilizing many different territorial- and community-based resources, inside a great variety of action model instruments and solutions, it appears particularly suitable for the application of tools enabling and making easier the direct participation of territorial stakeholders in the design and implementation of new innovative actions. The contribution of scientists and researchers is here related to the efforts in translating complexity and scientific discussion into available good quality knowledge contents integrated into more easy-to-access tools. Tools are intended for different types of territorial stakeholders: (1) active subjects in the governance and funding activities for fostering resilience practices, (2) active subjects in the operational design of the individual practices, and (3) potential target and end direct beneficiaries of the practices results and outputs.

14.6 Learning from practices and back to action—the role of tools

Practices stabilization, considering all the meaning of this concept explored previously, and scale-up efforts are for sure a crucial step toward transition to more resilient territories and communities—a transition in which sustainable development and climate change mitigation and adaptation issues are strictly coupled with resilience matters.

Such a working framework facilitates the boosting of social and technical innovation in the available toolboxes while reducing disciplinary and methodological boundaries over time. These produce dynamic actions and policy environments in which renewed goals for resilience enhancement not only tend to improve the capabilities and resources to react, in a positive way, to existing shocks and crises in specific fields and sectors, but also act in a proactive perspective.

In this reasoning, scaling up from individual practices or actions programs also means to support the development of models and methodologies for learning lessons from the field. Tools are available which enable and enhance the learning processes from operational experiences in spaces characterized by exchanges, sharing, and coproduction, where scientific research actors, experts, and practitioners meet territorial and community actors and stakeholders.

The measurement and assessment of resilience results is also a challenging topic. The identification of assessment indicators and measure units might allow better quality evaluations and comparison over time and among different territorial spaces. Measures are needed, together with the understanding of practices development, to boost governance, organizational, and funding resources. On the other hand, such requirements might produce "temptations of standardization" that could reduce the range of options and creativity to those identified as the reference elements for measurement. This risk is already visible, in some ways, in sustainable development, greening and circularity in the field of economic practices, and the evaluation of the commitments and performance of public and private bodies. The aim is not, of course, to question the crucial importance of measurements and assessments as such but to highlight a potential risk. The search for "the best indicators" means to identify and select measures describing, in general terms, (1) the topics and aims of the undertaken action; (2) the action goals, taking into consideration that the presence or absence itself of practices can be considered itself an indicator in this respect; (3) the operational approaches and the technical matters which characterize the practice feasibility and implementability, the functioning of the activities to be put in place, the system of involved active actors, and target subjects and expected results; (4) define the evaluation elements and thresholds on which basis the practice can be considered achieved and successful. The problem therefore becomes that of developing assessment tools which enable the understanding of effectiveness and efficiency in achieving the direct and indirect expected results but, on the other hand, risk to reduce the range of possible creative and innovative proposals because standards to not consider them. This is, for instance, the case of taxonomies and standards which are periodically being discussed at the international level about the certification of sustainable finance initiatives, circular and green economy activities and of the financial reporting reflecting the concrete and effective green-oriented action and performance of economic subjects in sustainability matters (see, among others, the activities of companies offering strategic planning, audit, tax, and advisory services to public and private bodies). The risk for the proliferation of "washing" practices, that is that of challenging and ambitious declarations of engagement which correspond to far below real activities, should also be faced with other tools, like the enhancement of the customer's awareness about what real practices should be.

Fostering resilience practices and territorial and community resilience in a stable and effective way for sure requires tools able to promote active involvement of all societal forces, innovation, and enhancement of learning and education processes.

References

Aubrey, K., & Riley, A. (2019). *Understanding and using educational theories* (2nd ed.). Sage Publisher.

Lave, J., & Wenger, E. (1991). *Situated learning: Legitimate peripheral participation*. Cambridge University Press.

Mejri, O., & Pesaro, G. (2015). Knowledge based strategies for disaster risk reduction: a knowledge management framework to increase understanding and awareness of the value of prevention and preparedness. In L. Podofillini, B. Sudret, B. Stojadinovic, E. Zio, & W. Kröger (Eds.), *Safety and reliability of complex engineered systems*. CRC Press Taylor and Francis Group.

Milliken, J. (2013). Resilience: From metaphor to an action plan for use in the peacebuilding field. In: *Geneva peacebuilding platform paper series, N.7/2013*.

Sharifi, A. (2016). A critical review of selected tools for assessing community resilience. *Ecological Indicators*, *69*, 629–647.

Wegener, E. (1998). *Communities of practice: Learning, meaning and identity*. Cambridge University Press.

Further reading

Baibarac, C., & Petrescu, D. (2019). Co-design and urban resilience: visioning tools for commoning resilience practices. *CoDesign*, *15*(2), 91–109.

Kachergis, E. J., Knapp, C. N., Fernandez-Gimenez, M. E., Ritten, J. P., Pritchett, J. G., Parsons, J., Hibbs, W., & Roath, R. (2013). Tools for resilience management: multidisciplinary development of state-and-transition models for northwest Colorado. *Ecology and Society*, *18*(4), 39. Available from https://doi.org/10.5751/ES-05805-180439.

CHAPTER 15

Creative diversity: facing Anthropocene challenges fostering resilience capacities

Angela Colucci
REsilienceLAB/Co.O.Pe.Ra.Te. Ldt, Italy

15.1 Enabling capacities: proprieties from resilience approaches

Aligning and establishing new cohesive approaches to face contemporary challenges is complex and arduous. The introduction underlines the tensions in translating the urban/territorial resilience concept into action. The research and literature debate often emphasizes the differences and gaps between approaches to resilience facing climate, social and economic stresses, risk (Brand & Jax, 2007; Burayidi et al., 2020; Davoudi et al., 2012), and unpredictable shocking events (as pandemic crisis due to COVID-19). The tensions between recovery, adaptation, and evolutionary resilience, between disciplines and models related to risk and climate change, between psychological and social resilience are the subject of a broad literature (Boschma, 2015; Pike et al., 2010).

Applying the concepts of persistence/recovery, adaptability, and transformability to community-led practices requires an adjustment of their definitions (Davoudi et al., 2013):

- persistence, when transition initiatives aim to cope with local urgencies related to social, economic, and environmental/climate impacts in the short term;
- adaptability, when transition initiatives aim to incremental adjustments facing local and regional/global phenomena;
- transformability, when transition initiatives experiment with innovative approaches in a long-term perspective able to promote changes at the local level, influencing and activating societal transition of the whole urban system.

[ECO]systems of Resilience Practices
DOI: https://doi.org/10.1016/B978-0-12-819198-9.00008-9

Resilience has become an often-overused umbrella concept, especially in media and policies discourses, emphasizing the positive envisioning potential of the concept itself (Colucci, 2012b; White, 2010).

Being aware of the importance of understanding and making explicit the gaps between disciplines to establish alliances, this chapter aims to deal with different approaches underlining common issues developed in cases and practices compared restarting from the properties that characterize the resilience concept in different disciplinary fields.

The entire book focuses on the transition initiatives and practices acting toward the resilience capacities enhancement. In the chapter the resilience practices are assumed as processes driven and acted by multiple actors' partnerships improving and modifying local urban context in their physical and functional/organizational components. The practices, simplifying their complex geography, act on commons: the interventions are oriented to improve public spaces, common environments, common goods and services through community-led actions generating positive benefits on environmental pressure phenomena and community strengthening. The energies and creativity emerging from the practices inspired the idea to highlight positive and common assets reframing the resilience properties and investigating promising crosscutting domains to foster intersectoral, multilevel, and multiactor alliances and synergies among community-led practices and cities' policies, programs, and projects.

Resilience, used more and more frequently in the debate on the innovation of urban regeneration models, is a concept of considerable richness enabling cultural innovation of urban and territorial design and management processes (Lerch, 2017; Nüchter et al., 2021). The success of resilience as a recurrent narrative metaphor is rooted in the capabilities in dealing with the multiple dimensions of recovery, transition, and adaptation time frames, with the multiple vulnerabilities of urban socio-ecological systems, often involving social equity and environmental sustainability questions (Carpenter et al., 2001; Galderisi & Colucci, 2018). The attention is shifted from resilience to resilience properties to reduce the vagueness, often characterizing the narrative of resilience, focusing on resilience as a set of capacities of complex systems, and investigating these capacities applied to complex urban/territorial systems (Page, 2011; Portugali & Stolk, 2016).

Different disciplines have used and use the resilience term by developing definitions related and functional to the research focus of each disciplinary field. In particular, in ecology and psychology, the approach to resilience solicits coping and learning to live with uncertainty (Adger, 2000): changes and crises are part of the evolutionary processes of complex systems. One of the key strategies to maintain and increase resilience mechanisms is perceiving dynamic conditions and grasping changes with adaptive responses oriented toward opportunities for growth and evolution.

In psychology, resilience defines the ability to cope with stressful or traumatic events and positively reorganize one's life in the face of difficulties (Chapman et al., 2018; Fletcher & Sarkar, 2013; Berkes & Ross, 2013). Therefore aspects and properties of the resilience of the individual are the ability to find "strength" to react (ability to react) in a positive way and the ability to learn from experiences, even traumatic (positive learning).

Ecological resilience is the property of ecosystems to react to stress phenomena, activating response, and adaptation strategies to restore the functioning mechanisms (Berkes et al., 2000). Resilient systems react by renewing themselves but maintaining the functionality and recognizability of the systems themselves in the face of stress.

Multiple properties contribute to the resilience of an ecosystem (Colucci, 2012a): creative diversity and redundancy, emergent properties, flexibility and modularity, functional mechanisms and renewal cycles, feedback cycles, and memory (some examples without the intention of a complete and comprehensive discussion and listing).

In physic and engineering, resilience is the specific propriety of materials that describes the capacity to bounce back to the initial condition after a deformation. Although the chapter does not focus on the contrasts between the disciplinary approaches, it is relevant to emphasize at least the dyscrasia concerning the assumption of multiple and dynamic equilibria instead of assuming one state of equilibrium (Folke & Gunderson, 2010; Holling, 1973).

Some properties and concepts are shared by all disciplinary definitions and approaches, such as flexibility and elasticity. Both materials and individuals and complex systems are more resilient if more elastic and flexible (e.g., they can react to "stresses" by storing and then returning energy). Other concepts are shared by the ecological, psychological, and social approaches, such as redundancy, creative diversity, learning/memory processes, feedback functionalities, and systemic organization (Elmqvist et al., 2003; Folke et al., 2004; Gunderson, 2000). The renewal cycles and evolutionary perspectives are concepts used in ecology and psychology. Renewal cycles are typical of ecosystems (Gunderson & Holling, 2002), but in the same way in psychology, the perspective of overcoming moments of crisis toward stages of renewed equilibrium that are dynamic and constantly in evolution is a fundamental concept for the psychological resilience. In the socio-ecosystems, the focus refers not only to the crisis/shock overcoming but also embrace long-term scenarios' definition: the aware and intentional design towards more desirable configurations (such as sustainability) Several properties empower the resilience capacities of socio-ecosystems (Berkes & Ross, 2013): the memory, learning capacities, and feedback loops characterize natural ecosystems, in which functional relationships and reactions are guided through the mechanisms of ecological memory, which intervenes in the composition and distribution of organisms and their interactions in space and time and storing the experience accumulated during the fluctuation of environmental conditions.

Feedback loops allow ecological systems to recognize signals of malfunction or disturbance and activate responses for adaptation based on learning and memory capacities. Learning from experiences and coming out stronger (positive coping) are essential aspects of individual psychological resilience and fundamental concepts when applied to territorial systems (Berkes et al., 2000).

15.2 Creative diversity: property for adaptive cities

As introduced earlier, the creative diversity and redundancy emerged for their potential in connecting multifaced and heterogeneous geographies of resilience in action. The narrative on commons or grassroots initiatives often emphasizes creativity as the essential factor enabling innovative experimentations (Weber, 2003). The presented panorama confirms the creative diversity as propriety connecting and navigating the tensions about resilience applied to complex territorial systems, and fostering alliances and synergies among disciplinary fields and across the policies sectors. Redundancy and creative diversity could be recognized as the immediate sources for replacing functions lost due to a disruptive event and provide the stock for activating adaptive responses concerning a multiplicity of temporal and spatial dimensions.

In ecology an ecosystem has three types of diversity: biological, genetic, and functional (Page, 2011). Each species is a single component of biodiversity itself, stored in the species' lineage (genetic diversity) and expressed through its behavior (functional diversity). In ecology, functional equivalence or functional redundancy is the ecological phenomenon in which multiple species representing various taxonomic groups may share similar, if not identical, roles in ecosystem function. Redundancy and creative diversity are insurance strategies for maintaining ecosystems' recognizability/functionality in the preferred regime threshold (Elmqvist et al., 2003).

Biodiversity is crucial in conferring adaptive capacity for renewal and reorganization: Gunderson (2000) argued that resilience is maintained by keystone structuring processes across spatial and temporal scales, sources of renewal and reorganization, and functional biodiversity. Elmqvist et al. (2003) and several studies on the natural ecosystems demonstrated that response diversity provides adaptive capacity and contributes to the resilience of desired ecosystem states (Nyström et al., 2000) and how functional response diversity, biological inheritance, mobile linkages, and ecological memory are the biological resources for adaptive renewal and reorganization after a disturbance.

The relevance of this propriety to the resilience of the socio-ecosystems is evidenced by the fact that the First Principle for Resilient Systems proposed by the Stockholm Resilience Centre (Biggs et al., 2015) is precisely that of maintaining diversity and redundancy: in the socio-ecological system, all components (species, landscape types, knowledge systems, actors, cultural groups, or institutions) offer different options in coping with uncertainty and shocks. The principle underlines how systems with differentiated components, including actors, or knowledge sources, are more resilient than simplified systems. In general, redundancy is confirmed as complex systems insurance compensating for the loss or failure of single components. In particular, functional diversity is crucial concerning response diversity (when components react differently to disturbances).

Creative diversity and redundancy are also strategic resources in psychology and sociology: social systems are, in general, more robust when multiple components participate. Research noticed how diversity (community participation) in decision-making processes improves innovative strategies, leading to better regimes for human–environment interaction (Liu et al., 2007; Ungar, 2018). Finally, diversity is one of the core properties of complex adaptive systems (Page, 2011).

15.3 Creative diversity for our common futures: resilience in public life

According to the emerging relevance of creative diversity in resilience practices exploration and Resilience Practices Observatory (RPO) project activities, the REsilienceLAB association promoted a coproduction path to address synergies and connections among resilience practices in action and the creative diversity soliciting the contribution and mutual exchanges of and among the different disciplinary approaches characterizing the association (ecology, landscape ecology, environmental planning/urban design, psychology and sociology, informatics and environmental engineering, economy). From the debate the need to transfer the theoretical stimuli operationally maintaining the complexity as a fundamental aspect in approaching multifaceted challenges that characterize our territories and urban systems emerged.

Assuming that resilience practices are acting on commons, the design of cities and territories in their public component emerges as the natural focus in retracing the resilient practices and creative diversity: increasing the creative and functional diversity of public urban components (spaces, functions, networks, nodes) at physical/spatial, immaterial, and virtual level appears as a fundamental strategy for strengthening the urban and territorial resilience (Colucci, 2018; Russegger et al., 2011). From the practices geographies the transversal and holistic project of public city/ public life) emerges as promising infrastructure strengthening the resilience of complex systems and ecological-environmental, social, economic, functional, and organizational components.

The public urban components, as a network of public spaces and public urban functions, emerge as strategic physical and immaterial infrastructure to strengthen the properties of creative and functional diversity of the local urban system.

The challenge is to foster the contribution of the practices in transforming cities into active agents of resilience and sustainability. Boosting creative diversity and redundancy of the public components of our cities enables the improvement of the preparedness to cope with unexpected shocks activating effective responses of recovery, allowing socio-ecosystems to adapt to changes (positive and negative), and to learn from past or ongoing crises and fostering the imaginative capacity of more sustainable and inclusive development models.

Based on the resilience practices discussed, the promising emerging action assets boosting the creative diversity of urban common/public components were clustered in four principal domains:

- space for nature: ecological diversity and evolution and functional diversity of green spaces for wellness,
- space for social diversity: the creativity of all social components and the functional creative diversity in public life,

- spaces for functional diversity: creative economic diversity and innovative urban functions, and
- organizational and creative diversity of governance processes and knowledge.

15.4 Creative diversity: actions domains for adaptive cities

The action domains traced emerge from the action focuses of resilience practices presented in the book, particularly from the urban/territorial components addressed. In this sense, the aim is to retrace practices action themes by emphasizing the domains on which the practices themselves contribute (or can contribute) to strengthening creative diversity and redundancy.

15.4.1 Space for nature: ecological biodiversity and human–natural health and well-being

Preserving and improving biodiversity is a frequent issue in resilience practices, both in communities- and institutional-led processes. The regional/territorial green infrastructure projects are principally planned/promoted by planning policies and implemented through territorial scales projects usually held by large partnerships and financed by national or European programs (Eggermont et al., 2015). The practices' focus is more at a local scale and oriented to urban/local biodiversity and on the paranatural habitat in artificialized contexts improvement.

Green public spaces constitute strategic "infrastructure" for the resilience of the city and the public city. Neo-ecosystems and Nature-Based Solutions (European Commission. Directorate General for Research & Innovation, 2015; European Environment Agency, 2021) have to be integrated into the network of green open spaces and components of the built environment. The practices provide examples in proposing various typologies of actions and several initiatives boost local communities' stewardship of urban green networks. For instance, the Re Lambro project activated a codesign process involving the Politecnico of Milan students and the local communities to develop design solutions and physically improve the ecological habitats along the Lambro River in Milan (Fig. 15.1). The process stimulated synergies activation among different local groups, allowing the realization of community gardens, ecosystem restoration, neo-habitats rehabilitation, and didactical exploration. The project activated a bottom-up process recognizing the Lambro river values

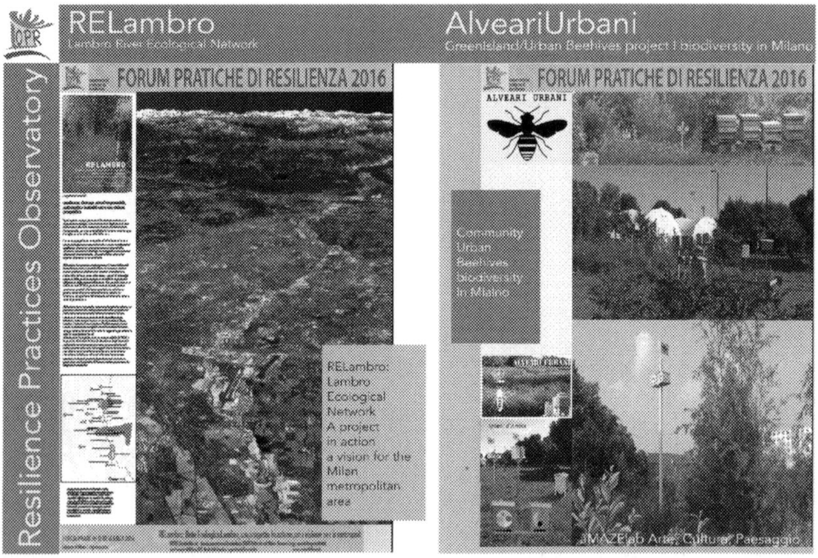

Figure 15.1 Resilience Practices Obeservatory practices: RE Lambro and Alveari Urbani projects (with the posters presented by practices for the Resilience Practice Forums). *Credit: Author elaboration based on the RPO documentation.*

(cultural, ecological, and in terms of services performed) and realising ecological restoration interventions, new ecosystems, and community gardens.

The design of urban (and territorial) green networks and infrastructures must include a strong differentiation of new green ecosystems, ensuring a diversity of habitats and microhabitats capable of performing different ecosystem services permeating all project components.

The reconstruction of different paranatural habitats in green open spaces can provide essential ecosystem services of regulation, support, supply, and cultural (Acuto & Leffel, 2021). At territorial or regional scales, green infrastructures connect parks and new green with large areas with large trees or urban forestation, green and blue connections, and core functional habitats (e.g., wetlands). In parallel, the local scale plays a vital role (Cortinovis & Geneletti, 2019): increasing and connecting all the small free spaces can improve the biodiversity of plant associations and habitats/ecosystems. For instance, flower meadows and non-invasive management systems can support essential ecosystem services such as pollination, enhancing the ecological mechanisms supporting the ecosystem services essential for urban climate adaptation (Pickett et al., 2013).

The local, including the micro-ones, interventions promoted by resilience practices and the interventions launched in peripheral urban areas or along blue corridors contribute to the ecological functionality restoration and improvement.

Practices are also acting on the improvement of urban and local biodiversity. For instance, several practices act on the pollination service improving the complexity and diversification of gardens, making them attractive for pollinating insects. Alveari Urbani (https://www.facebook.com/claudiazanfi.green/, Fig. 15.1), in Milano, promoted the urban beehives realized with local and international artists and managed by local communities in urban parks. The projects contribute to the beehives' installation, beekeeping activities, and the management of flower and herbaceous gardens that are appetible for insects. Several RPO practices developed local interventions in restoring small habitats to preserve wild species living in urban or periurban areas.

An extensive range of practices encouraged (and encourages) creative natural and ecological diversity, implementing/realizing complex and multiscale projects involving communities and institutional academic stakeholders. These practices provided two indirect benefits that can also be underlined as promising perspectives in practices contribution: promoting the mainstreaming of nature-based solutions and ecosystem services as common ground for all practices transformations and fostering the synergies and arrangements among local-territorial scales and research/community actors both crucial tfor green infrastructures and nature-based solutions implementation.

Local resilience practices are often solicited to be consistent with long-term policies and programs operating at regional and European levels. At the same time, local initiatives demonstrate strong empathy and connection among local communities and urban nature diversity, promoting a profound rebalance and recognition of ecological features (e.g., the Transition Network and other international initiatives foster the knowledge and the implementation of agroecological and biological agricultural models).

During the pandemic event the green places and their diversity became vital resources for human and social well-being. Human well-being cannot be separated from the heath of the planet and the natural and ecological components (principle assumed in the One-Health approach introduced by World Health Organization, https://www.who.int/news-room/q-a-detail/one-health) (Rabinowitz et al., 2018).

Green infrastructures are therefore a strategic tool for well-being of communities: in absorbing pollutants and improving the microclimate outdoor comfort, in offering services of psychological benefit, furthermore they are places to stimulate healthy and therapeutic activities (Barp et al., 2015; Cushing & Miller, 2020).

Green and collective forms of green spaces management are interventions often launched by resilience practices: several innovative forms of urban community gardens, social farming, and social agriculture demonstrated how the management of green spaces also became an opportunity for well-being enhancement of involved local communities and vulnerable/fragile groups through training and stewardship activities. Practices also often propose innovative therapeutic uses of urban green spaces, where green spaces accommodate functions and activities that can be conducted outdoors and in connection with places of social activation (learning, knowledge sharing) such as sports activities and courses not necessarily all related to traditional facilities and equipment. A very stimulating example is the projects promoted by Associazione Professionale Italiana Danzamovimentoterapia (APID, Italian chapter of European Association Dance Movement Therapy, https://www.apid.it/danzamovimentoterapia/) that launched dance movement therapy paths in spaces (also green/open) placed in central Italy small villages affected by recent earthquakes to activate processes of community and identity re-building. In some practices the realization of community gardens and shared green areas constitutes physical interventions able to strengthen community building and promote well-being and therapeutic paths for vulnerable/fragilepopulations.

15.4.2 Spaces for social diversity and innovation

Chapter 1 underlines the relevance of including human diversity in the resilience practices and transitional processes. Universal inclusion, regardless of all the personal conditions, is crucial in enabling and engaging all citizens in resilience practices, amplifying resources and stimuli for innovation and evolution.

The common public city must enable everyone to contribute. Everyone's diversity and uniqueness are indispensable resources for change: interspecies diversity in nature is a vital asset: it is essential to grasp its importance in our society. The design of the public city must be based on the assumption that diversity is a resource for renewal and nurturing imaginative capacities for possible futures envisioning. The different components (spaces, functions, physical and virtual organizations) of the public city must enable everyone to contribute to strengthening the resilience of the complex system (Borowczyk, 2018; Persson et al., 2015). Diversity means enabling everyone; furthermore, it means to share commitments and responsibilities in local transitional projects toward a more livable, sustainable, and resilient urban and territorial contexts.

A crucial suggestion that arose from the RPO practices is the need to enhance the inclusion and diversity concerning the emergency procedures and preparedness, enabling and making everyone aware of risks and able to react during emergencies and critical situations. In particular, the "Children, are we prepared against environmental disasters?" promoted by the Gli Altri ed i Disastri association (www.glialtriedidisastri.org) aimed to focus on immigrants, urban risks, and emergency. In partnership with local cultural associations, the project organized workshops (performance and games) targeted at children to improve awareness about the territorial risk characterizing the local context. The project faced three crucial criticalities: enable immigrants' children and their families to understand the risks, enable immigrants to access/understand the early warning and emergency procedures and promoting individual responsiveness to emergencies enabling immigrant communities to be active and aware actors of emergency management (self-help).

The practices and projects on the public spaces/public life propose nodes to strengthen social relationships and embrace diversified social networks intercepting multiple geometries from micro-social networks of inhabitants to international networks able to activate generative and creative processes of sharing and exchange (initiatives such as the re-activation of places and spaces for co-production of knowledge, for exchange of experiences and skills, and local projects co-design).

The design of spaces and places that accommodate multiple activities and services is crucial to guarantee the differentiation of functions and the redundancy of functions improving the resilience capacities of socio-ecological systems. The initiatives increasing and qualifying the spaces dedicated to collective activities, contribute to diversified empowered communities. Likewise, activating seeds of cultural creative and collaborative functions is a vital driver of innovation, with positive effects on the resilience of the social-ecological systems. Several resilience practices experimented (and experiment) with innovative models for community facilities and public function in urban and internal areas. Educational and skill-sharing are very diffused and inspiration sources also for public and institutional-led policies. Projects, as SCHOOL.01 (https://www.school01.org, Fig. 15.2) are promoting multiple uses of public buildings, integrating, for example, art exhibition and lab in public school, therefore promoting social inclusion and social community building. It is possible to mention also projects as Progetto Scuola Villaggio (school-village project, Fig. 15.2), proposing alternative and integrative learning activities to support the regeneration of internal areas (as well as other projects focus on health-care or other welfare services integrated by community/sharing models).

15.4.3 Functional creative diversity: economic and urban models innovation

Resilient complex systems also need diversified economic components: the diversification of productions, enterprise models, and products guarantees adaptive capacities in overcoming local and global crises affecting climate, social and economic components. Innovative economic models rooted in sustainable principles are today a growing phenomenon and trend involving different sectors from the more conventional ones to innovation and research companies, including diffusion of enterprises that include the production of social and

Figure 15.2 Resilience Practices Observatory practices: School.01 and Scuola Villaggio projects (with the posters presented by practices for the Resilience Practice Forums). *Credit: Author elaboration based on the RPO documentation.*

environmental benefits to the local system in their business mission. In particular, innovative forms of social enterprise can be a strategic resource in the management of services to the community, contributing, in addition to direct benefits, to the strengthening of the functional redundancy of the local urban system, multiplying the capacity to offer strategic services such as services related to education, care, and assistance, but also in the local management of environmental cycles and services such as water and waste, integrating services already offered in an often centralized way. Enhancing local system resilience is essential to support innovative green economy initiatives that arise from collaboration among multiple actors and seeing a redistribution of benefits as a prerequisite for their success (energy communities, circular economies, and bioeconomics). The practices are an incredible source of innovation and imaginative solutions, from the more consolidated practices based on circular and blue economies to more innovative ideas based on community multiple benefits or sharing models. Several practices promote new functions for abandoned forests to restore the landscape and ecological quality, one example is the Boschi Vivi project, which proposed an innovative "diffuse" cemetery model (Boschi Vivi—https://boschivivi.it/, Fig. 15.3) activating the restoration of forest or the GALLAB project that regenerated traditional abandoned artisanal activities

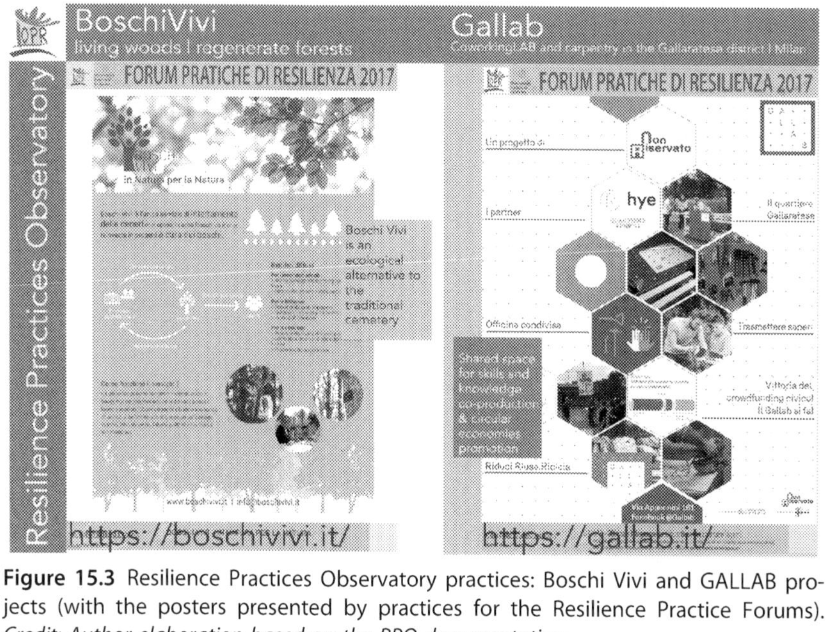

Figure 15.3 Resilience Practices Observatory practices: Boschi Vivi and GALLAB projects (with the posters presented by practices for the Resilience Practice Forums). *Credit: Author elaboration based on the RPO documentation.*

(reactivating both the places and the culture) opening shared spaces where helder and young generations share skills to launch new social entrepreneurial activities based on circular economy principles (GALLAB—https://gallab.it/, Fig. 15.3).

15.4.4 Organizational creative diversity: innovating governance processes and knowledge

The creative diversity in the management of complex governance processes and the hybridization of knowledge are crucial for promoting the cultural and sociocultural transition toward more resilient systems.

Creative diversity inspires the comparison, dialogue, and integration among different epistemologies (hard sciences, sciences and humanities, arts, and material knowledge) toward alliances and cultural advancements. Knowledge hybridization can create the conditions to improve cultural diversity and enable cultural transition toward more resilient societies.

Fostering alliances, as boosted during the Resilience Practices Forum labs events (Chapter 5) among academic/researchers, associations, and communities active on social inclusion, and networks on creativity and arts to deal with climate challenges and imagining solutions permits to arose innovative and successful projects and to reach remarkable impact on the citizen, also in terms of awareness and topics diffusion. The practices also demonstrated the relevance of virtual and physical media and solutions to improve the attention and the awareness on sustainability through local festivals and events, organizing training from citizens to citizens, involving performing arts (as forum theater, concerts events, exhibitions), or through more virtual media and tools exploiting the potential of information and communications technology.

Creativity and arts demonstrated their incredible power in supporting imaginative paths for territorial and urban regeneration and reactivation. A large number of practices, like Nature, Art & Habitat Residency (https://nahr.it/) or Favara cultural Farm (https://www.farmculturalpark.com/), showed as starting from a private initiative promoting performing arts and cultural activities as levers for reactivating dismissed or abandoned internal areas, it is possible to activate successful long-term process engaging all local and territorial actors and communities able to provide regenerated local economies, identities, and values. The storytelling of the practices (presented by people involved and committed) proved to be a powerful and successful instrument to share the resilience practices principles and diffuse the sustainability and resilience issues awareness.

The diversity of stakeholders, the redundancy of their roles, and their active participation in processes improve the resilience of social and ecological systems, making natural and social environments more sustainable. The adaptive governance model is aimed at accompanying the process of urban transformation, strengthening and diversifying the organizational components (Bixler et al., 2020; Folke et al., 2005; Ostrom, 1990; Voss et al., 2006). The involvement of different actors by scale, skills in the construction phase, implementation, and management of public city projects ensures the resilience of the overall system. It increases the feasibility of the transformation, reduction of "costs" (distributed among actors) in the management phase of projects of social interest and the public city, and strengthening of awareness and the role that local communities, associations, and citizens can have in improving the quality of the public city (Borgström, 2019). Chapter 16 focuses on governance dilemmas and challenges enabling the resilience practices stabilization as agents of sustainability transitions.

Acknowledgments

The author is grateful to all the REsilienceLAB members for their contributions and inspirational discourses during the Creative Diversity workshops paths.

References

Acuto, M., & Leffel, B. (2021). Understanding the global ecosystem of city networks. *Urban Studies, 58*(9), 1758–1774. Available from https://doi.org/10.1177/0042098020929261.

Adger, W. N. (2000). Social and ecological resilience: Are they related? *Progress in Human Geography, 24*(3), 347–364. Available from https://doi.org/10.1191/030913200701540465.

Barp, A., Bolla, D., & Vittadini, M. (2015). *Spazi verdi da vivere: Il verde fa bene alla salute*. Il Prato.

Berkes, F., Folke, C., & Colding, J. (2000). *Linking social and ecological systems: Management practices and social mechanisms for building resilience*. Cambridge University Press.

Berkes, F., & Ross, H. (2013). Community resilience: Toward an integrated approach. *Society and Natural Resources, 26*(1), 5–20. Available from https://doi.org/10.1080/08941920.2012.736605.

Biggs, R., Schlüter, M., & Schoon, M. L. (Eds.), (2015). *Principles for building resilience: Sustaining ecosystem services in social-ecological systems*. Cambridge University Press.

Bixler, R. P., Lieberknecht, K., Atshan, S., Zutz, C. P., Richter, S. M., & Belaire, J. A. (2020). Reframing urban governance for resilience implementation: The role of network closure and other insights from a network approach. *Cities (London, England), 103*102726. Available from https://doi.org/10.1016/j.cities.2020.102726.

Borgström, S. (2019). Balancing diversity and connectivity in multi-level governance settings for urban transformative capacity. *Ambio*, *48*(5), 463–477. Available from https://doi.org/10.1007/s13280-018-01142-1.

Borowczyk, J. (2018). Sustainable urban development: Spatial analyses as novel tools for planning a universally designed city. *Sustainability*, *10*(5), 1407. Available from https://doi.org/10.3390/su10051407.

Boschma, R. (2015). Towards an evolutionary perspective on regional resilience. *Regional Studies*, *49*(5), 733–751. Available from https://doi.org/10.1080/00343404.2014.959481.

Brand, F. S., & Jax, K. (2007). Focusing the meaning(s) of resilience: resilience as a descriptive concept and a boundary object. *Ecology and Society*, *12*(1). Available from https://www.jstor.org/stable/26267855.

Burayidi, M. A., Allen, A., Twigg, J., & Wamsler, C. (2020). *The Routledge handbook of urban resilience*.

Carpenter, S., Walker, B., Anderies, J. M., & Abel, N. (2001). From metaphor to measurement: Resilience of what to what? *Ecosystems*, *4*(8), 765–781. Available from https://doi.org/10.1007/s10021-001-0045-9.

Chapman, D. A., Trott, C. D., Silka, L., Lickel, B., & Clayton, S. (2018). Psychological perspectives on community resilience and climate change. In *Psychology and Climate Change* (pp. 267–288). Elsevier. Available from https://doi.org/10.1016/B978-0-12-813130-5.00011-4.

Colucci, A. (2012a). *Le città resilienti: Approcci e strategie*. Università, Polo interregionale di eccellenza Jean Monnet.

Colucci, A. (2012b). Towards resilient cities. Comparing approaches/strategies. *Tema. Journal of Land Use, Mobility and Environment*. Available from https://doi.org/10.6092/1970-9870/921, 101-116 Paginazione.

Colucci, A. (2018). *Transition initiatives: Three exploration paths. Smart, resilient and transition cities: Emerging approaches and tools for a climate-sensitive urban development*. Elsevier.

Cortinovis, C., & Geneletti, D. (2019). A framework to explore the effects of urban planning decisions on regulating ecosystem services in cities. *Ecosystem Services*, *38*100946. Available from https://doi.org/10.1016/j.ecoser.2019.100946.

Cushing, D. F., & Miller, E. (2020). *Creating great places: Evidence-based urban design for health and wellbeing*. Routledge: Taylor & Francis Group.

Davoudi, S., Brooks, E., & Mehmood, A. (2013). Evolutionary resilience and strategies for climate adaptation. *Planning Practice and Research*, *28*(3), 307–322. Available from https://doi.org/10.1080/02697459.2013.787695.

Davoudi, S., Shaw, K., Haider, L. J., Quinlan, A. E., Peterson, G. D., Wilkinson, C., Fünfgeld, H., McEvoy, D., Porter, L., & Davoudi, S. (2012). Resilience: A bridging concept or a dead end? "reframing" resilience: Challenges for planning theory and practice interacting traps: Resilience assessment of a pasture management system in Northern Afghanistan urban resilience: What does it mean in planning practice? Resilience as a useful concept for climate change adaptation? The politics of resilience for planning: A cautionary note. *Planning Theory & Practice*, *13*(2), 299–333. Available from https://doi.org/10.1080/14649357.2012.677124.

Eggermont, H., Balian, E., Azevedo, J. M. N., Beumer, V., Brodin, T., Claudet, J., Fady, B., Grube, M., Keune, H., Lamarque, P., Reuter, K., Smith, M., van Ham, C., Weisser, W. W., & Le Roux, X. (2015). Nature-based solutions: New influence for environmental management and research in Europe. *GAIA - Ecological Perspectives for Science and Society*, *24* (4), 243–248. Available from https://doi.org/10.14512/gaia.24.4.9.

Elmqvist, T., Folke, C., Nyström, M., Peterson, G., Bengtsson, J., Walker, B., & Norberg, J. (2003). Response diversity, ecosystem change, and resilience. *Frontiers in Ecology and the Environment*, *1*(9), 488–494. Available from https://doi.org/10.1890/1540-9295(2003)001[0488:RDECAR]2.0.CO;2.

European Commission. Directorate General for Research and Innovation. (2015). *Towards an EU research and innovation policy agenda for nature-based solutions & re-naturing cities. Final report of the Horizon 2020 expert group on 'Nature based solutions and re naturing cities': (full version)*. Publications Office. Available from https://data.europa.eu/doi/10.2777/765301.

European Environment Agency. (2021). *Nature-based solutions in Europe policy, knowledge and practice for climate change adaptation and disaster risk reduction*. Publications Office. Available from https://data.europa.eu/doi/10.2800/919315.

Fletcher, D., & Sarkar, M. (2013). Psychological resilience: A review and critique of definitions, concepts, and theory. *European Psychologist, 18*(1), 12–23. Available from https://doi.org/10.1027/1016-9040/a000124.

Folke, C., Carpenter, S., Walker, B., Scheffer, M., Elmqvist, T., Gunderson, L., & Holling, C. S. (2004). Regime shifts, resilience, and biodiversity in ecosystem management. *Annual Review of Ecology, Evolution, and Systematics, 35*(1), 557–581. Available from https://doi.org/10.1146/annurev.ecolsys.35.021103.105711.

Folke, C., & Gunderson, L. (2010). Resilience and global sustainability. *Ecology and Society, 15*(4). Available from https://doi.org/10.5751/ES-03954-150443, art43.

Folke, C., Hahn, T., Olsson, P., & Norberg, J. (2005). Adaptive governance of social-ecological systems. *Annual Review of Environment and Resources, 30*(1), 441–473. Available from https://doi.org/10.1146/annurev.energy.30.050504.144511.

Galderisi, A., & Colucci, A. (2018). *Smart, resilient and transition cities: Emerging approaches and tools for a climate-sensitive urban development*. Elsevier.

Gunderson, L. H. (2000). Ecological resilience—In theory and application. *Annual Review of Ecology and Systematics, 31*(1), 425–439. Available from https://doi.org/10.1146/annurev.ecolsys.31.1.425.

Gunderson, L. H., & Holling, C. S. (Eds.), (2002). *Panarchy: Understanding transformations in human and natural systems*. Island Press.

Holling, C. S. (1973). Resilience and stability of ecological systems. *Annual Review of Ecology and Systematics, 4*(1), 1–23. Available from https://doi.org/10.1146/annurev.es.04.110173.000245.

Lerch, D. (Ed.), (2017). *The community resilience reader: Essential resources for an era of upheaval*. Island Press.

Liu, J., Dietz, T., Carpenter, S. R., Alberti, M., Folke, C., Moran, E., Pell, A. N., Deadman, P., Kratz, T., Lubchenco, J., Ostrom, E., Ouyang, Z., Provencher, W., Redman, C. L., Schneider, S. H., & Taylor, W. W. (2007). Complexity of coupled human and natural systems. *Science (New York, N.Y.), 317*(5844), 1513–1516. Available from https://doi.org/10.1126/science.1144004.

Nüchter, V., Abson, D. J., von Wehrden, H., & Engler, J.-O. (2021). The concept of resilience in recent sustainability research. *Sustainability, 13*(5), 2735. Available from https://doi.org/10.3390/su13052735.

Nyström, M., Folke, C., & Moberg, F. (2000). Coral reef disturbance and resilience in a human-dominated environment. *Trends in Ecology & Evolution, 15*(10), 413–417. Available from https://doi.org/10.1016/S0169-5347(00)01948-0.

Ostrom, E. (1990). *Governing the commons: The evolution of institutions for collective action* (1st ed.). Cambridge University Press. Available from https://doi.org/10.1017/CBO9780511807763.

Page, S. E. (2011). *Diversity and complexity*. Princeton University Press.

Persson, H., Åhman, H., Yngling, A. A., & Gulliksen, J. (2015). Universal design, inclusive design, accessible design, design for all: Different concepts—One goal? On the concept of accessibility—Historical, methodological and philosophical aspects. *Universal Access in the Information Society, 14*(4), 505–526. Available from https://doi.org/10.1007/s10209-014-0358-z.

Pickett, S. T. A., Cadenasso, M. L., & McGrath, B. (Eds.), (2013). *Resilience in ecology and urban design* (Vol. 3). Netherlands: Springer. Available from https://doi.org/10.1007/978-94-007-5341-9.

Pike, A., Dawley, S., & Tomaney, J. (2010). Resilience, adaptation and adaptability. *Cambridge Journal of Regions, Economy and Society*, *3*(1), 59–70. Available from https://doi.org/10.1093/cjres/rsq001.

Portugali, J., & Stolk, E. (Eds.), (2016). *Complexity, cognition, urban planning and design*. Springer International Publishing. Available from https://doi.org/10.1007/978-3-319-32653-5.

Rabinowitz, P. M., Pappaioanou, M., Bardosh, K. L., & Conti, L. (2018). A planetary vision for one health. *BMJ Global Health*, *3*(5), e001137. Available from https://doi.org/10.1136/bmjgh-2018-001137.

Russegger, G., Tarasiewicz, M., & Wlodkowski, M. (Eds.), (2011). *Coded cultures: Creative practices out of diversity*. Springer.

Ungar, M. (2018). Systemic resilience: Principles and processes for a science of change in contexts of adversity. *Ecology and Society*, *23*(4). Available from https://doi.org/10.5751/ES-10385-230434, art34.

Voss, J.-P., Bauknecht, D., & Kemp, R. (Eds.), (2006). *Reflexive governance for sustainable development*. Edward Elgar.

Weber, E. P. (2003). *Bringing society back. Grassroots ecosystem management, accountability, and sustainable communities*. MIT Press.

White, I. (2010). *Water and the city: Risk, resilience and planning for a sustainable future*. Routledge.

CHAPTER 16

Organizational (creative) diversity. Dilemmas and perspectives of governance processes for complex systems resilience

Angela Colucci
REsilienceLAB/Co.O.Pe.Ra.Te. Ldt, Italy

16.1 Resilience practices: heterogeneous geographies and potentialities

Since the last decades, academic and disciplinary debate have increasingly paid attention to the heterogeneous geographies of resilience practices, also named transition initiatives, spontaneous interventions, and grassroots innovations (Fayaz Ahmad Sheikh & Bhaduri, 2021; Hossain, 2016; Moallemi et al., 2020).

The increased attention belongs to the inspirational resources of experimented innovations of urban and territorial transformation and the objective growth, pervasiveness, and spread of these heterogeneous practices. Differentiated initiatives activated and managed by local groups acting on both physical dimension (mainly on urban public space) and immaterial dimension as environmental issues or social services and facilities of commons became a relevant worldwide phenomenon (Smith et al., 2014).

In planning, urban design and policies debate their role became more central (Banerjee, 2001). Several projects propose renovated approaches to urban transformation, integrating and valorizing energies from spontaneous initiatives promoting participative and inclusive codesign processes. It is possible to collect a wide range of initiatives that launched and managed experiments of urban regeneration processes innovations both emphasizing the direct engagement of citizens and communities and enabling alliances among multiple experts/local communities actors acting directly on

[ECO]systems of Resilience Practices
DOI: https://doi.org/10.1016/B978-0-12-819198-9.00016-8

urban physical and immaterial innovation of commons: it can be cited the Handmade Urbanism experience (Rosa & Weiland, 2013), in which the urban design and regeneration process are directly acted and managed by citizens engaged since the decision-making process through coproduction and codesign tools, or the Tactical Urbanism experience where urban designers are also directly actors of urban interventions (Lydon et al., 2015). The attention paid to spontaneous initiatives from the academic debate is not recent; the US pavilion presented in the 2012 Biennale di Venezia the event on "Spontaneous Interventions: Design Actions for the Common Good" that celebrated and presented initiatives of citizens able to transform critialities and urgencies of urban context into new opportunities (http://www.spontaneousinterventions.org/) and demonstrated a consolidated exploration and recognition of the values and contribution of spontaneous practices.

In literature debate, transition initiatives are investigated through a differentiated lens concerning innovative and creative proposals and solutions (grassroots innovations, social innovation), community-led governance processes (coproduction and codesign, participative urban interventions), and social aspects (commons and public services, urban and public policies) (Colucci, 2018b; Smith et al., 2014). The book, as well as this chapter, assumes an inclusive definition of resilience practice as a process driven by mixed partnerships involving local communities activating an extensive range of action domains promoting local transformation and providing multiple benefits in contributing to the long-term transitions towards a more robust, aware, and adaptive society and sustainable territorial systems. In literature, under the transition initiatives (Frantzeskaki et al., 2012), umbrella is included a broad range of local/community-based processes led by citizen groups (non-governmental organization (NGOs)) with private and expert actors and acting on different sets of issues from social innovation and inclusion to sustainable local behaviour and collectively addressing the sustainability goals (Colucci, 2017). A peculiarity is the differentiated range of issues addressed that usually result from a balanced fluid equilibrium among local urgencies characterizing the local context, priorities of actions defined by local actors, and territorial or global long-term visions (such as sustainable development goals, urban or metropolitan climate policies, or social program). This fluid balancing process often enables unexpected and creative solutions and activates experimental interventions characterizing every practice by a unique narrative history, even if similar initial conditions characterize them. Even in the cases in which the levers of local initiatives are rooted in social urgencies (for instance, social innovation initiatives), environmental and climate change issues are frequently integrated into transition initiatives.

Some commonalities emerge from the practices and initiatives exploration (Chapter 5), allowing a common characterization of resilience practices as initiatives local-based, community-led, crosscutting/multi-issues, enabling the experimentation of innovative solutions, coping with urban urgencies, and based on inclusive governance processes.

16.2 Good governance/adaptive governance: urgency and crosscutting issues

The term governance refers to all the mechanisms through which individuals, organizations, and institutions (public and private) collaborate in managing common interests and social affairs (Duit et al., 2010; Salamon & Elliott, 2002), including self-regulation or informal rules of civil society, coregulation of public and private actors, and regulation through institutional government. In these terms, governance refers to structures and processes that allow social organizations to make decisions and share power enabling the conditions of collective action and organizational social coordination (Duit et al., 2010; Folke et al., 2005; Folke, 2007; Migaud et al., 2021; Olsson et al., 2006).

Concerning territorial and urban physical and immaterial transformations, the governance process has been intended as the whole process starting from the decision-making, including problem setting and the development of strategic visions, to the implementation and management of policies, measures, and actions/interventions (Colucci, 2018a).

Governance emerges as one of the main barriers to stabilizing and implementing community-led initiatives (Chapter 5 and 8). This chapter emphasizes the urgent need for advancements in conceptual reframing and governance models to better deal with the complexity and interconnected effects among environmental, social, economic, and economic organizational phenomena.

Several authors (Dietz, 2003; Liu et al., 2007; Ostrom, 2004) denounced the need for a general renovation of governance processes proposing alternative/innovative models: models such as adaptive governance (Folke, 2007) or reflexive governance (Maesschalck, 2017; Voss et al., 2006) embrace cross-scale interactions embedding the multiscale phenomena and dynamics of complex systems. Furthermore, the need in renovating governance represents an explicit goal of international and regional programs: Sustainable Development Goal number 17 refers to the need to "Strengthen the means of implementation and revitalize the global partnership for sustainable development," several UN-HABITAT campaigns promote the Good Urban Governance (https://mirror.unhabitat.org/content.asp?typeid = 19&catid = 25&cid = 2097) (New Urban Agenda, 2017; UN-Habitat, 2016) and the improvement of territorial/urban governance is the core issue and challenge of interregional EU policies (Knieling & Leal Filho, 2013). The shift from current governance models toward more adaptive (Charalabidis & Koussouris, 2012) and collaborative models is a common challenge for cities all over the world, being cities increasingly called to integrate emerging community-led initiatives and institutional climate strategies into the broader process of sustainable development of complex urban systems (Batliwala, 2002; Boyle & Kay, 2001).

The fundamental reason for which all disciplines, from social and political sciences to more technical disciplines, address the governance dimension in a more or less explicit way is that governance refers to all the aspects enabling the implementation and management of policies, programs, projects, and actions/interventions at local, territorial, national, and international/global levels. The process governance intercepts all the transformations toward sustainability and resilience proposed and promoted by different sectors and actors.

The governance, especially when referred to resilience practices, emerges as a wicked problem. The practices actors (as in the case of the practices of the Resilience Practices Observatory) report difficulties and criticalities in the management of the implementation processes: therefore a clear perception of the need to deploy skills and tools to improve the quality of the processes appears but at the same time, the actors often vaguely refer to the governance of processes only in terms of "criticalities/barriers concerning the realization of specific actions/solutions." Understanding the different levels and complexities of the governance process is crucial to consciously design the process by identifying solutions and using appropriate tools to assure the implementation and achieve targeted results.

In approaching this wicked challenge related to the resilience practices governance, the discourse could be scattered along the scales to address the different questions and outline possible solutions:

- The micro or endogenous scale refers to the quality and robustness of the process of the single practice.
- The meso or endogenous scale includes the questions on networking among practices and the ones on integration with the urban or local territorial system.
- The macroscale refers to the theoretical and ethical questions emerging in assuming the resilience practices as an actor in the sustainable global goals achievement.

16.3 Micro or endogenous governance: the design of practices process

Micro or endogenous governance refers to how the single resilience practice approaches and manages the process of their initiatives locally. Numerous organizational barriers emerge from the case studies presented. Common barriers to numerous practices are related to the difficulties in citizens' engagement and participation and maintaining active and alive initiatives while activating a stabilization of process able to germinate new initiatives and actions. The RPO experience has highlighted how the actors promoting practices are aware of these weaknesses and difficulties. At the same time, it emerges a lack of a conscious and intentional project of the governance process: which actors to involve, which tools to guarantee an open decision-making process, how to achieve the participation and involvement of direct beneficiaries, how to extend the benefits to local communities.

16.3.1 Inclusiveness in practices governance process

Internal decision-making process is a crucial aspect not explicitly addressed in (and by) resilience practices. Foremost, this aspect could be referred to as the inclusiveness of the practices' governance process. It is common to assume that it is "per se" inclusive if the community leads a practice. As the resilience practices exploration revealed, the initiatives' internal decision-making process is often, but not always inclusive and participative. Unfortunately, in some cases, the leading actors manage the practice without an inclusive involvement of local communities' actors.

The result in some cases is that the solutions and actions, mainly if transition initiatives are acting on public spaces or public services, are established without a participative and inclusive process involving all the interested citizens and stakeholders. The lack of participation often induces difficulties and conflicts in projects' implementation or even a process of gentrification of some (undesired) sectors of users and citizens. An indirect impact of the lack of inclusive process is also a scarce interest and participation of citizens even though they were the primary beneficiaries of the initiatives themselves.

The exploration of RPO experiences revealed that resilience practices are mainly based on open governance process of codesign and large/mixed partnerships, but this attitude often is not an intentional precondition or principle in practice design. From the literature on the Transition Network (Feola & Nunes, 2014; Smith, 2011) and the RPO experience, it arises that large and inclusive partnerships are among the main factors for the success of transition initiatives. The RPO experience demonstrates how resilience practices based on mixed partnerships, including institutions, research and expert actors, private/economic actors, and communities/NGOs organizations, have more chances of evolving in a consolidated process with significant positive impacts and contributions toward local systems sustainability.

The partnership is in the majority mixed, but the inclusiveness is often based on local conditions and dynamics of collaborations already existing among actors and not established by an aware and explicit investigation of existing stakeholders, community groups, or associations concerning different action domains.

Making the process inclusive and participative could be a way to guarantee democracy in practices decision-making and a crucial challenge for the practices stabilization and for valorizing and multiplying the benefits. Tools as stakeholders and local community maps can support a more inclusive identification of actors and local community interests to be engaged in the practice decision-making process.

Monitoring and self-evaluation tools can also support the accountability and transparency of resilience practices governance process. In resilience practices cases, some monitoring and assessment tools were developed concerning specific issues. However, the monitoring and assessment of the process is not often proposed/developed in community-led initiatives. In the Transition Network experience, practices are solicited to celebrate efforts and reflect (Smith, 2011): these two tools can contribute to transparency and evaluation of the initiative process celebrating results reached and reflecting on results and barriers). The networking and the involvement/engagement are principal factors in evaluating the success of the local transition initiative of Transition Initiatives (https://transitionnetwork.org/) (Hopkins, 2008; Transition Network team, 2016).

The explicit activation of strategies and tools for citizen involvement, improving the transparency and inclusiveness of the decision-making process, can support the overcoming of engagement difficulties. The Transition Town Network, for example, developed tools and training modules supporting transition initiatives in improving local capabilities to engage citizens, ensuring the growth and stabilization of the process.

16.3.2 Island effect

At the same time the explorations presented show a certain degree of fragility of local transition initiatives also when they are part of a more comprehensive network: in several cases the actors promoting the initiative noticed as they feel isolated. This island effect could be related to the difficulty of a single local initiative to be effectively connected with institutions and organizations acting at a more extensive (urban) scale.

The result is the flourishing of numerous initiatives, the duplication/overlay of networks, and, meanwhile, the lack of dialog, synergies, and alliances activation among them. This aspect is even more crucial in climate/resilience issues that require integrated and intersectoral strategies and actions.

16.4 Meso or exogenous governance: the practices networking

Thanks to their nature and abilities in engaging local communities and implementing solutions and experimentations on the fields, resilience practices became actors in territorial institutional-led processes for climate change policies mitigation and adaptation and urban resilience improvement. In the last two decades an extensive range of metropolitan/urban policies involved and engaged transition initiatives in long terms and strategic processes on climate change and sustainability-related issues (energy, mobility, and food policies) (Acuto & Leffel, 2021). The transition initiatives became crucial actors both in the decision-making process and in policies implementation.

Literature on Transition Towns Network investigates the internal debate about the Transition Towns movement role in the local arena (Barnes, 2015; Feola & Butt, 2017; MacKinnon & Derickson, 2013): the Network assumed to allow/support partnerships arrangements and cooperation with institutions/government only if the local Transition Towns Initiative maintains a recognizable autonomy in the polices governance processes.

16.4.1 Networks of resilience practices

The networking that could be officially established or an informal cooperation among association/practices emerges as promising organizational structure providing capacities building, training, and cultural instruments (recognition and identity) to support local community-led/spontaneous interventions. The networking organizational structure is crucial in promoting the development of long-term and more complex strategic visions related, for example, to urban climate issues, low carbon, or environmental and social sustainability. The resilience practices network enables the conditions (knowledge, skills, organizational, recognizability) to stabilize and upscale community-led initiatives (Acuto & Leffel, 2021; Goh, 2020).

Networks and alliances of resilience practices are activated for different reasons and present different characteristics and structures (Galderisi & Colucci, 2018):

- networks focusing on shared issues/aims or specific/thematic action domain (such as food, climate action, social innovation); these networks usually develop tools for mutual support (expert contributions, knowledge sharing, solutions sharing) and are characterized by a horizontal structure coordinating local initiatives;
- networks sharing principles and aims referred to global issues, as the Transition Town Network; these networks are more structured and provide to local practices long-term vision, training, and supporting tools; networking is based on shared values and vision and is usually characterized by a more robust organizational structure;
- networks of local/regional practices addressing similar topics and issues, but strictly limited to a metropolitan or regional context: these networks promote knowledge sharing, mutual support for initiatives, and their partnerships often include also local institutions.

For resilience practices, networking is a promising organizational structure enabling synergies with institutions and private actors. It is also a promising and innovative organizational structure for the governance framework supporting the urban complex system transition processes and enabling the resilience practices contributions.

Networking structures can support the stabilization, upscaling, and diffusion of resilience practices: based on the first successful practices experiment(s) the networks often activate a dynamic process of establishment of main actions principles (ethical values, set of "global" goals) and toolboxes development (how to start and how to do). As for the Transition Network, the citizens' groups that share principles and long-term visions can join the network to be supported in transition initiative activation and management. Since the beginning, the Transition Network has demonstrated a relevant commitment in the network organizational structure definition, in shared long-term vision/principles establishment and, at the same time, providing identity (feeling to be part of a global movement) and supporting (training, toolboxes) single Transition Towns initiatives.

16.4.2 Practices networking: enabling potentialities

The resilience practices networking supports the partnerships among actors at different scales enabling the multiscale dimension of transition initiatives otherwise embedded at the local scale. Likewise, it facilitates the alliances with other community-led practices acting on similar topics, aims, and perspectives or focusing on similar tools action domains offering a framework for potential upscaling (Ravi, 2019).

The resilience practices networks have developed tools for learning and knowledge coproduction/codesign that are fundamental tools for sharing knowledge resources, identifying criticalities and challenges to face, and defining priorities of local actions and initiatives (Amin & Cohendet, 2004; Berkes, Colding, et al., 2000).

Knowledge coproduction and training are fundamental to valorize and capitalize local community knowledge resources and sources from expert knowledge to operative and action-oriented skills, to engage all local community components in sharing their competencies, knowledge, and expertise, and to amplify practices contribution generating a shift of cultural sphere toward more sustainable individual and social behaviors (Minkler et al., 2008).

Resilience practices networks can also support local initiatives' awareness of their contribution to global phenomena, such as climate change mitigation and adaptation. Hence, local initiatives could intercept and dialog with the existing public policies framework. Networks among local practices can give a larger scale or multiscale perspective to local practices, providing advantages in terms of constancy and complexity both to the local context of the community-led practice and the larger urban/regional context.

16.5 Macro level: governance of transition processes

16.5.1 Governance of transition processes

The transition management focus is the governance of transition processes and how, concerning established long-term sustainability visions, innovative transition initiatives can influence modifications of meso level toward more sustainable states. The transition management governance model proposes a radical shift from conventional models based on decision-making (goals and priorities) and implementation (actions) oriented to achieving a set of targeted goals in a defined time to adaptive and dynamic governance models. These shifts emphasize stakeholders' engagement in establishing multiple visions articulated in alternative scenarios to reach and shared set of goals. At the same time, opportunities and obstacles to transformations could emerge from experiments developing technological and social innovations that play a crucial role in transition management for future development trajectories outlining. Strong connections between long and medium terms (meso-level regime and macro-level regime) characterize this transition model, connecting vision, and short-term action through circular and iterative processes (Loorbach, 2010). Four governance activities (Loorbach et al., 2017) promoting societal transitions characterize the transition arena: strategic activities focusing on system transition in a long-term period (30 years), tactical activities focusing on subsystem transition in medium-term (5–15 years), operational activities focusing on practices (project and actions) in short-term period, and reflexive activities. Reflexive activities refer to continuous monitoring as a vital part of transitions' research and learning process (Taanman et al., 2008) to monitor transition experiments in particular, concerning social and institutional learning level, and to evaluate the entire transition process (progress, barriers, and opportunities). Kemp et al. (2007), Rotmans and Loorbach (2009), and Rotmans et al. (2001) applied the transition concept to complex systems concerning sustainable development, activating transitions processes of structural change in specific societal domains such as energy supply, housing, mobility, agriculture, healthcare (Geels, 2002).

In socio-technological transition management literature, monitoring the transition process is crucial in reorienting policies and measures and supporting fertile environments for innovation niches. Monitoring activities and tools could be efficacious instruments in documenting local multiple benefits and providing tangible data to assess the transition initiatives' contribution to local and global sustainability.

16.6 Adaptive governance: resilience practices contribution and governance dilemmas

The need for a profound renovation of governance processes clearly and urgently emerges requiring a radical shift from current governance models, generally based on the linear and consequential process starting from goal definition, policies prioritization to implementation phase toward adaptive governance models (Olsson et al., 2004) capable of engaging different stakeholders, emphasizing their role in setting multiple visions structured according to alternative scenarios aiming the achievement of a shared and constantly updated set of goals (Rotmans & Loorbach, 2009).

The concept of "strategic long-term scenario" proposed by transition management approaches has similarities with the threshold/boundaries of the socio-ecosystems concept (Berkes, Folke, et al., 2000; Folke et al., 2004; Walker et al., 2004): the long-term strategic scenario presents conceptual linkages with the establishment of the preferred regime threshold for socio-ecological systems adaptation and evolution dynamics. Both the long-term strategic vision and the socio-ecological system preferable threshold are based on dynamic equilibria. Both leave space and opportunities for multiple possible trajectories of local system dynamics (including resilience practices) of evolution, which are necessary to cope with constant changes in environmental and economic conditions (Kemp et al., 1998).

In this framework the networking of transition initiatives emerges as a promising organizational structure that, if effectively integrated into adaptive urban governance framework, could allow linking the different niches of innovation and the more structured and consolidated organizational structures and stakeholders (institutional, civil society, private actors, universities), and it could become an innovative space of dialog among different stakeholders (Bodin, 2017; Charalabidis & Koussouris, 2012; Folke, 2007). By including the network of transition initiatives in the adaptive urban governance framework, it could be possible to promote and support the stabilization, upscaling, germination, and improvement of community-led and transitions initiatives, improving the creative diversity and redundancy proprieties of urban systems and sustaining a large catchment of innovation sources for adaptation. At the same time the resilience practices and their networking increase the organizational diversity of territorial/urban transformation processes, improving the systems' resilience capacities (see also Chapter 17, which discuss the perspectives of [eco]system of resilience practices as an innovative organizational structure).

A conflictual dilemma emerges correlating the transition management model and the Ostrom discourses on institutional diversity and governance, recognizing the practices acting on commons as a third part autonomous and separate from institution and market. In the societal transition toward sustainable scenarios, the "innovation niches" and resilience practices are actors/players in/of a larger multilevel and multiactor processes frameworks. In the project implemented, applying the transitional governance models, the governance of large process appears intentionally designed (energy transition, for example) based on multistakeholder decision-making process and initiatives innovating and promoting local and small experimentation are encouraged. The societal transitional governance models are rooted in long-term visions established by multistakeholder and dynamic decision-making processes in which the three parts (civil society/commons, institutions, and market) are all involved. According to the Intuitional Diversity (Ostrom, 2005) theoretical framework, it emphasizes the relative independence of the "resilience practices" not necessarily embedded and concurring toward predefined long-term visions/scenarios.

Considering the EU context, the dilemma could remain on the level of theoretical speculation, without having significant impacts on operative processes of territorial system transformation: the institutions, the market, and the civic society (in which for simplicity they include here the practices of resilience) act in a common framework of democracy and respect for common values and rights (McGinnis & Ostrom, 2014; Ostrom, 1990, 2004). Although these speculations are not a focus of this chapter, it is essential to remark how the dilemma can envisage ethical implications. Defining the communities-led as a third party or involved actors in defining and reaching shared scenarios between actors is not only theoretical thinking referring to the contexts where there is no substantial democracy and freedom (Balamir, 2021), and where community practices are often in contrast with national or long-term regional visions. In these terms the resilience practices can address climate change challenges or sustainable development goals globally shared but avoided (or repressed) by national or regional governments.

Otherwise, it is not assumable di per sè that each resilience practice assumes and advocates ethical principles (democracy of process, social inclusion). The networking/[eco]system of resilience practices can foster the inclusiveness and democracy of the practice governance process (Creech et al., 2008; Hughes et al., 2020; Turnhout et al., 2010).

Both networking cases presented adopted tools to guarantee the assumption of essential values: in the RPO the resilience practices definition also includes selection criteria, and in Transition Network the practices adhesion is based on sharing the Network values and principles. The networking structures of resilience practices could foster awareness and the adoption of accountability tools to guarantee (or boost) a more transparent and explicit assumption and integration of ethical/democracy principles in the practices governance processes (Baird et al., 2016; Weber, 2003). Independently from the democracy dilemma discourse, more transparent, open, and accountable decision-making processes appear powerful and promising solutions for practices stabilization to enhance the recognition of resilience practices as crucial and essential actors toward sustainability and resilience capabilities increase of territorial systems.

References

Acuto, M., & Leffel, B. (2021). Understanding the global ecosystem of city networks. *Urban Studies, 58*(9), 1758–1774. Available from https://doi.org/10.1177/0042098020929261.

Amin, A., & Cohendet, P. (2004). *Architectures of knowledge: Firms, capabilities, and communities*. Oxford University Press.

Baird, J., Plummer, R., & Bodin, Ö. (2016). Collaborative governance for climate change adaptation in Canada: Experimenting with adaptive co-management. *Regional Environmental Change, 16*(3), 747–758. Available from https://doi.org/10.1007/s10113-015-0790-5.

Balamir, S. (2021). *Unsustaining the commodity-machine: Commoning practices in postcapitalist design* [PhD thesis, Amsterdam School for Cultural Analysis (ASCA), Faculty of Humanities (FGw)]. https://dare.uva.nl/search?identifier = 41641ebe-104c-4572-9b5f-56f64c9390a1.

Banerjee, T. (2001). The future of public space: Beyond invented streets and reinvented places. *Journal of the American Planning Association, 67*(1), 9–24. Available from https://doi.org/10.1080/01944360108976352.

Barnes, H. (2015). Exploring the factors that influence nurse practitioner role transition. *The Journal for Nurse Practitioners, 11*(2), 178–183. Available from https://doi.org/10.1016/j.nurpra.2014.11.004.

Batliwala, S. (2002). Grassroots movements as transnational actors: Implications for global civil society. *Voluntas: International Journal of Voluntary and Nonprofit Organizations, 13*(4), 393–409. Available from https://doi.org/10.1023/A:1022014127214.

Berkes, F., Colding, J., & Folke, C. (2000). Rediscovery of traditional ecological knowledge as adaptive management. *Ecological Applications, 10*(5), 1251–1262. Available from https://doi.org/10.1890/1051-0761(2000)010[1251:ROTEKA]2.0.CO;2.

Berkes, F., Folke, C., & Colding, J. (2000). *Linking Social and Ecological Systems: Management Practices and Social Mechanisms for Building Resilience*. Cambridge University Press.

Bodin, Ö. (2017). Collaborative environmental governance: Achieving collective action in social-ecological systems. Science, *357*(6352), eaan1114. Available from https://doi.org/10.1126/science.aan1114

Boyle, M., & Kay, J. J. (2001). *Monitoring in support of policy: An adaptive ecosystem approach*. 48.

Charalabidis, Y., & Koussouris, S. (Eds.), (2012). *Empowering Open and Collaborative Governance*. Berlin Heidelberg: Springer. Available from https://doi.org/10.1007/978-3-642-27219-6.

Colucci, A. (2018a). *The Transition approach in urban innovations. Smart, Resilient and Transition Cities* (pp. 19–28). Elsevier. Available from https://doi.org/10.1016/B978-0-12-811477-3.00003-1.

Colucci, A. (2018b). *Transition initiatives: Three exploration paths. Smart, Resilient and Transition Cities: Emerging Approaches and Tools for A Climate-Sensitive Urban Development*. Elsevier.

Colucci. (2017). Resilience injections in urban public realm design. Maggioli.

Creech, H., Vetter, T., Matus, K. J. M., & Seymour, I. R. (2008). *The governance of non-legal entities: An exploration into the challenges facing collaborative, multistakeholder enterprises that are hosted by institutions* [Monograph]. International Institute for Sustainable Development. http://www.iisd.org/publications/pub.aspx?id = 1044.

Dietz, T. (2003). The struggle to govern the commons. *Science*, *302*(5652), 1907–1912. Available from https://doi.org/10.1126/science.1091015.

Duit, A., Galaz, V., Eckerberg, K., & Ebbesson, J. (2010). Governance, complexity, and resilience. *Global Environmental Change*, *20*(3), 363–368. Available from https://doi.org/10.1016/j.gloenvcha.2010.04.006.

Fayaz Ahmad Sheikh., & Bhaduri, S. (2021). Policy space for informal sector grassroots innovations: Towards a 'bottom-up' narrative. *International Development Planning Review*, *43*(1), 115–137. Available from https://doi.org/10.3828/idpr.2019.34.

Feola, G., & Butt, A. (2017). The diffusion of grassroots innovations for sustainability in Italy and Great Britain: An exploratory spatial data analysis: The diffusion of grassroots innovations for sustainability in Italy and Great Britain. *The Geographical Journal*, *183*(1), 16–33. Available from https://doi.org/10.1111/geoj.12153.

Feola, G., & Nunes, R. (2014). Success and failure of grassroots innovations for addressing climate change: The case of the Transition Movement. *Global Environmental Change*, *24*, 232–250. Available from https://doi.org/10.1016/j.gloenvcha.2013.11.011.

Folke, C. (2007). Social–ecological systems and adaptive governance of the commons. *Ecological Research*, *22*(1), 14–15. Available from https://doi.org/10.1007/s11284-006-0074-0.

Folke, C., Carpenter, S., Walker, B., Scheffer, M., Elmqvist, T., Gunderson, L., & Holling, C. S. (2004). Regime shifts, resilience, and biodiversity in ecosystem management. *Annual Review of Ecology, Evolution, and Systematics*, *35*(1), 557–581. Available from https://doi.org/10.1146/annurev.ecolsys.35.021103.105711.

Folke, C., Hahn, T., Olsson, P., & Norberg, J. (2005). Adaptive governance of social-ecological systems. *Annual Review of Environment and Resources*, *30*(1), 441–473. Available from https://doi.org/10.1146/annurev.energy.30.050504.144511.

Frantzeskaki, N., Loorbach, D., & Meadowcroft, J. (2012). Governing societal transitions to sustainability. *International Journal of Sustainable Development*, *15*(1–2), 19. Available from https://doi.org/10.1504/IJSD.2012.044032.

Galderisi, A., & Colucci, A. (2018). *Smart, Resilient and Transition Cities: Emerging Approaches and Tools for a Climate-Sensitive Urban Development*. Elsevier.

Geels, F. W. (2002). Technological transitions as evolutionary reconfiguration processes: A multi-level perspective and a case-study. *Research Policy*, *31*(8–9), 1257–1274. Available from https://doi.org/10.1016/S0048-7333(02)00062-8.

Goh, K. (2020). Flows in formation: The global-urban networks of climate change adaptation. *Urban Studies*, *57*(11), 2222–2240. Available from https://doi.org/10.1177/0042098018807306.

Hopkins, R. (2008). *The Transition Handbook: From Oil Dependency to Local Resilience*. Green Books.

Hossain, M. (2016). Grassroots innovation: A systematic review of two decades of research. *Journal of Cleaner Production*, *137*, 973–981. Available from https://doi.org/10.1016/j.jclepro.2016.07.140.

Hughes, S., Giest, S., & Tozer, L. (2020). Accountability and data-driven urban climate governance. *Nature Climate Change*, *10*(12), 1085–1090. Available from https://doi.org/10.1038/s41558-020-00953-z.

Kemp, R., Loorbach, D., & Rotmans, J. (2007). Transition management as a model for managing processes of co-evolution towards sustainable development. *International Journal of Sustainable Development & World Ecology*, *14*(1), 78–91. Available from https://doi.org/10.1080/13504500709469709.

Kemp, R., Schot, J., & Hoogma, R. (1998). Regime shifts to sustainability through processes of niche formation: The approach of strategic niche management. *Technology Analysis & Strategic Management*, *10*(2), 175–198. Available from https://doi.org/10.1080/09537329808524310.

Knieling, J., & Leal Filho, W. (Eds.), (2013). *Climate Change Governance* (1. ed). Springer.

Liu, J., Dietz, T., Carpenter, S. R., Alberti, M., Folke, C., Moran, E., Pell, A. N., Deadman, P., Kratz, T., Lubchenco, J., Ostrom, E., Ouyang, Z., Provencher, W., Redman, C. L., Schneider, S. H., & Taylor, W. W. (2007). Complexity of coupled human and natural systems. *Science*, *317*(5844), 1513–1516. Available from https://doi.org/10.1126/science.1144004.

Loorbach, D. (2010). Transition management for sustainable development: A prescriptive, complexity-based governance framework. *Governance*, *23*(1), 161–183. Available from https://doi.org/10.1111/j.1468-0491.2009.01471.x.

Loorbach, D., Frantzeskaki, N., & Avelino, F. (2017). Sustainability transitions research: Transforming science and practice for societal change. *Annual Review of Environment and Resources*, *42*(1), 599–626. Available from https://doi.org/10.1146/annurev-environ-102014-021340.

Lydon, M., Garcia, A., & Duany, A. (2015). *Tactical Urbanism: Short-Term Action For Long-Term Change*. Island Press.

MacKinnon, D., & Derickson, K. D. (2013). From resilience to resourcefulness: A critique of resilience policy and activism. *Progress in Human Geography*, *37*(2), 253–270. Available from https://doi.org/10.1177/0309132512454775.

Maesschalck, M. (2017). *Reflexive governance for research and innovative knowledge. ISTE*. Wiley.

McGinnis, M. D., & Ostrom, E. (2014). Social-ecological system framework: Initial changes and continuing challenges. Ecology and Society, *19*(2), art30. Available from https://doi.org/10.5751/ES-06387-190230

Migaud, M. R., Greer, R. A., & Bullock, J. B. (2021). Developing an adaptive space governance framework. *Space Policy*, *55*, 101400. Available from https://doi.org/10.1016/j.spacepol.2020.101400.

Minkler, M., Vásquez, V. B., Tajik, M., & Petersen, D. (2008). Promoting environmental justice through community-based participatory research: The role of community and partnership capacity. *Health Education & Behavior*, *35*(1), 119–137. Available from https://doi.org/10.1177/1090198106287692.

Moallemi, E. A., Malekpour, S., Hadjikakou, M., Raven, R., Szetey, K., Ningrum, D., Dhiaulhaq, A., & Bryan, B. A. (2020). Achieving the sustainable development goals requires transdisciplinary innovation at the local scale. *One Earth*, *3*(3), 300–313. Available from https://doi.org/10.1016/j.oneear.2020.08.006.

New urban agenda: H III: Habitat III: Quito 17–20 October 2016. (2017). United Nations Conference on Housing and Sustainable Urban Development, Nairobi. United Nations.

Olsson, P., Folke, C., & Berkes, F. (2004). Adaptive comanagement for building resilience in social? Ecological systems. *Environmental Management*, *34*(1). Available from https://doi.org/10.1007/s00267-003-0101-7.

Olsson, P., Gunderson, L. H., Carpenter, S. R., Ryan, P., Lebel, L., Folke, C., & Holling, C. S. (2006). Shooting the rapids: Navigating transitions to adaptive governance of social-ecological systems. *Ecology and Society*, *11*(1), 18.

Ostrom, E. (1990). *Governing the Commons: The Evolution of Institutions for Collective Action* (1st ed.). Cambridge University Press. Available from https://doi.org/10.1017/CBO9780511807763.

Ostrom, E. (2004). Panarchy: Understanding transformations in human and natural systems. *Ecological Economics*, *49*(4), 488–491. Available from https://doi.org/10.1016/j.ecolecon.2004.01.010.

Ostrom, E. (2005). *Understanding Institutional Diversity*. Princeton University Press.

Ravi, R. V. (2019). *Partnership Among Grassroots Organizations*. MJP Publisher.

Rosa, M. L., & Weiland, U. E. (Eds.), (2013). Handmade Urbanism: From Community Initiatives to Participatory Models *(livre + DVD)*. Jovis Verlag GmbH.

Rotmans, J., & Loorbach, D. (2009). Complexity and transition management. *Journal of Industrial Ecology*, *13*(2), 184–196. Available from https://doi.org/10.1111/j.1530-9290.2009.00116.x.

Rotmans, J., Kemp, R., & van Asselt, M. (2001). More evolution than revolution:. *Foresight*, *3*(1), 15–31. Available from https://doi.org/10.1108/14636680110803003.

Salamon, L. M., & Elliott, O. V. (Eds.), (2002). *The Tools of Government: A Guide to the New Governance*. Oxford University Press.

Smith, A. (2011). The Transition Town Network: A review of current evolutions and renaissance. *Social Movement Studies*, *10*(1), 99–105. Available from https://doi.org/10.1080/14742837.2011.545229.

Smith, A., Fressoli, M., & Thomas, H. (2014). Grassroots innovation movements: challenges and contributions. *Journal of Cleaner Production*, *63*, 114–124. Available from https://doi.org/10.1016/j.jclepro.2012.12.025.

Taanman, M., de Groot, A., Kemp, R., & Verspagen, B. (2008). Diffusion Paths for Micro Cogeneration Using Hydrogen in the Netherlands. *Journal of Cleaner Production*, *16*(1), 124–132. Available from https://doi.org/10.1016/j.jclepro.2007.10.010.

Transition Network team. (2016). *The Essential Guide to Doing Transition. Getting Transition Started in Your Street, Community, Town or Organisation*. Transition Network. https://transitionnetwork.org/resources/essential-guide-to-transition-v-1/.

Turnhout, E., Van Bommel, S., & Aarts, N. (2010). How participation creates citizens: participatory governance as performative practice. Ecology and Society, *15*(4), art26. Available from https://doi.org/10.5751/ES-03701-150426

UN-Habitat. (2016). Urbanization and Development: Emerging Futures. UN-Habitat.

Voss, J.-P., Bauknecht, D., & Kemp, R. (Eds.), (2006). *Reflexive Governance for Sustainable Development*. Edward Elgar.

Walker, B., Holling, C. S., Carpenter, S. R., & Kinzig, A. P. (2004). Resilience, adaptability and transformability in social-ecological systems. Ecology and Society, *9*(2), art5. Available from https://doi.org/10.5751/ES-00650-090205

Weber, E. P. (2003). *Bringing Society Back In: Grassroots Ecosystem Management, Accountability, and Sustainable Communities*. MIT Press.

CHAPTER 17

Challenges for sustainability: fostering [eco]systems of resilience practices contribution

Angela Colucci
ResilienceLAB/Co.O.Pe.Ra.Te. Ldt, Italy

17.1 Resilience practices contributing to the commons: challenges and advancements

Ostrom (Ostrom, 1990, 2005) affirmed that self-organized systems have frequently solved many common dilemmas. The book chapters presented the Resilience Practices Observatory (RPO) project and several international experiences demonstrating the values of collective processes for sustainable long-term visions and confirming the increased relevance of these processes at the international level. In the last decades, the stakeholders of civil society and informal processes have increased their presence and engagement in the forums and platforms for global policies framework definition (for instance, the UN-HABITAT and United Nations Office for Disaster Risk Reduction (UNDRR) campaigns, the United Nations Framework Convention on Climate Change and in the framework of the Conference of Parties (COP) in particular since the Paris COP 21).

At the same time, the challenges of recent years demonstrated the inadequacy of a purely rational approach typically assumed by institutional policies in missing the core point of complexity and uncertainty. Often institutional policies, based mainly on calculative rationality, assume that the future could be assumed as knowable concerning given parameters (Batliwala, 2002). However, overcoming multiple priorities and emergencies from climate change, economic, and social crisis requires to include uncertainty and relative adaptability in the decision-making process and the implementation phases of policies (Roelich & Giesekam, 2019).

The resilience practices confirmed how practices shifted the approach: assuming the uncertainties and the indetermination of the future and acting in the present (Batliwala, 2002), experimenting with the feasibility of

[ECO]systems of Resilience Practices
DOI: https://doi.org/10.1016/B978-0-12-819198-9.00019-3

innovations, and implementing possible alternative models. The provision of creative innovations and the experimentation of alternative models represent a crucial contribution to the shift toward more sustainable models and territorial/urban complex systems. This chapter aims to discuss the advancements (or confirmed criticalities) along with the threats for common action and network stabilization highlighted by Ostrom (2005).

17.1.1 Rapid exogenous changes

This threat envisages the risk of inadequacy to respond to rapid local and global changes bringing the community-led process to substantial failure. The fragilities of the resilience practices and transition initiatives are underlined by several authors and emerged as a shared criticality about which the actors leading the resilience practices are aware, as noticed in Chapter 5 concerning the case of Resilience Practices Observatory.

The discourses collected in the Foreword about the resilience practices and the implications during the pandemic event of Covid-19 showed, from one side, the positive reaction of communities-led practices in adapting and reorganizing their activities to intercept and cope with the urgencies supporting local communities during the crisis. Form the other side the pandemic period showed a diffuse suspension of the activities launched by practices.

As discussed in Chapter 15 concerning the creative diversity propriety, it is essential to arrange the argumentation from respect to the two levels: the single resilience practice and the [eco]system of practices. If the single practice considered locally and isolated could be fragile concerning different aspects such as isolation, lack of robust governance design, the ecosystem of practices (or networking of) could be considered resilient based on a robust creative diversity and redundancy of the practices' system itself (Bixler et al., 2020).

Indeed, the fragility and the risk of failure due to not adapting to rapid changes in the context remain a risk associated with a single initiative. This threat is intrinsic in the fact that every single practice constitutes a seed and experimentation of innovation of urban spaces uses/functions, of services, of economies and businesses, of organization and partnership, and it implies a risk of failing (Loorbach et al., 2017). In this sense, maintaining and supporting innovation niches is essential in the sociocultural transition models (Colucci, 2018; Feola & Butt, 2017; Kemp et al., 2007).

At the same time, considering not the single practice but the system of practices, an advance is undoubtedly visible. The presence, role, and visibility of the movements promoted by civil society have grown. Worldwide networks such as Transition Networks show an overall advance in the stabilization and capacity to react and reorganize strategies, tools, and priorities to emergencies and changes (Acuto & Leffel, 2021; Batliwala, 2002; Ravi, 2019). Sharing and networking platforms are more and more numerous, leading to a slow stabilization of the [eco]system of practices and, as indirect effects, the single practice results more supported and the role and contribution of practices acting on commons become strengthened and recognized.

Moreover, the temporal dimension could constitute a further aspect for reflection: resilience practices, like all community-led initiatives, are characterized by long times period necessary in the early stages of launching the initiatives. A recurring aspect in the narration of the experiences of practices and initiatives concerns the long gestation times between the maturation of the idea and the launch of the first initiatives and the implementation of the first actions. Extended temporality is necessary for participatory and codesign processes which, however, turn out to guarantee greater robustness of practices in the long term: his time-period is often denounced as "lost" by actors but it is an investment for the construction of open and inclusive decision-making processes and often constitutes a factor of stabilization.

17.1.2 Transmission failure

This threat refers to the lack of adequate stabilization of the practice and the lack in the transfer of skills and knowledge necessary in the long-term management of activated processes that often implies a renovation in actors' coordinating/managing the process. This criticality can derive from incorrect management of the initiative process's governance and the lack of investment in the capacitation of actors delegated to manage the practice over time.

The intentional design of the governance process includes the identification of solutions to ensure its continuity and stabilization. Solutions such as the enlargement of the actors involved and engaged in managing the activated process and implementing open and participatory processes constitute effective strategies to avoid the risk of a failure or cessation of processes due to the lack of skills transmission.

The practices explored have shown, albeit still timid, to be aware of the crucial relevance and centrality of an open and inclusive process and the need to design the governance and management aspects. This awareness constitutes the first aspect of advancement. As demonstrated in Chapter 7, the practices address these weaknesses activating paths of knowledge sharing and solutions exchange.

Significant progress is indeed achieved concerning the involved actors' capacitation and strengthening the skills of coordinators and managers of the practice process.

In exploring the RPO project, the domain of capacity-building tools emerges as the most widespread and implemented by practices: the domain of the capacity-building tools (32%) represents the most adopted, or at least the most implemented/launched in the first steps of the practices' processes. The first step emerged as phase devoted to improving the capacities of the association/partners managing the practices and local communities or targeted beneficiaries.

In the first phase, capacity-building tools are mainly addressed in improving the skills and capacities of actors committed to managing resilience practices. The relevance and importance ascribed to the capacities improvement by the resilience practices could be considered an advancement reached.

The awareness and diffuse use of capacity-building tools could be assumed to be an advancement in facing the transmission failure threat, improving the actors' capacities increase the robustness and cohesiveness of organizations and partnerships that are more able to overcome and stabilize the process in the long-term perspective. The analysis concerning the germination and stabilization of the RPO practices demonstrated that 47% of resilience practices certainly germinated other synergic initiatives with the initial one, demonstrating stabilization of the transitional process.

17.1.3 Blueprint thinking

This threat regards replicating similar solutions unplaced with a substantial disconnection with local characteristics, priorities, and resources. The risk of failure concerns the disconnection and the low integration "in" and "with" the local context dynamics and features and the inadequacy in responding/facing local urgencies and priorities. The RPO project exploration reveals as a prevailing rootedness in the local context characterizes the resilience practices. Indeed, the initiatives or actions are launched as responses to the urgencies that arose from local contexts characterization. For example, under the general criticality of environmental phenomena and climate change (clusters and generalizations established to provide a synthetic interpretation of the RPO practices' geographies), the described phenomena of criticalities were adequately detailed, circumscribed, and connected with the local context and, in a second phase, also often linked to critical issues addressed by European or international policies (in particular the seventeen UN Sustainable Developemnte Goals) (United Nations General Assembly, 2015).

Furthermore, the identification of solutions proposed by resilience practices of the RPO project is commonly rooted in the local resources emphasizing the unicity and specificity of local contexts (cultural, landscape, social, economic). For these reasons the blueprint threat did not emerge as a risk characterizing the resilience practices of the RPO project. Concerning networking, the Transition Network (Chapter 11) shows similar dynamics: it implies a general adhesion to shared principles and provides guidelines/toolkits supporting single and local transition initiative activation, stabilization, and upscaling. As demonstrated by the Spanish case, the Transition Network discourages the solutions replication, and the toolkits developed focus on the process activation and management and are principally finalized to the capacitation in community building and engagement. The ingredients for prioritizing solutions and initiatives are rooted in local urgencies and community resources and priorities.

Blueprint copies or the replication of solutions are threats that could characterize initiatives connected to the thematic networks or could emerge in the mainstreaming and diffusion of some specific technical solution becoming a "model to be replicated" without an accurate contextualization. For example, numerous thematic networks have been activated in the last decades, supporting sectorial aspects of urban regeneration processes. In particular, some of these thematic networks are often highly specialized and sectorial (such as networking on urban public green management, green rooftop, green facades), having as principal focus sharing technical intervention or solutions.

The originality and specificity of resilience practices are favorable conditions in avoiding the blueprint threat. The originality is often a way to recognize and manifest a specific identity of the resilience practice (the actors are proud about the unicity of the practices). A more diffusion of the transversal toolkit, like in the case of Transition Network, or mutual capacitation and narrative support in solving processual and emergent criticalities demonstrate, more than a threat, a perspective for the [eco]systems of practices stabilization and advancement.

17.1.4 Corruption

The corruption issue did not emerge in any phase or material of the RPO project, except for some specific and very limited mentions of corruption phenomena in some urban and territorial policies/transformations. This positive absence is due primarily to the fact that the practices are mainly proposed and managed by associations and groups of citizens whose efforts (of energy, skills, and time) are not even comparable with the sources of funding and support. Even the dimension of corruption regarding the "powers," which is an aspect that could characterize, in general, the processes of practices, did not emerge in the case of RPO project practices.

The chapter on governance (Chapter 16) debates the need for more accountability and transparency of the governance processes, which could also be assumed to be a crucial tool in preventing corruption affecting more easily the initiatives with closed process and rooted in small and homogeneous partnerships (Hughes et al., 2020; Weber, 2003).

17.1.5 Lack of large-scale arrangements

This phenomenon is closely connected to the tension, frequently discussed in the Contributions, between processes often firmly rooted in the local context and the need to contribute and be integrated into large-scale and long-term views and policies.

The resilience practices exploration confirms that the local dimension is the prevailing scale of intervention. For sure, it is related to the community dimension of the initiative itself: communities are acting on local contexts, activating different projects based mainly on volunteers and on citizen engagement aiming to give impact and tangible results on local context such as the improvement of urban environments or urban functions and services. Furthermore, the community-led experiences are characterized by profund relationships among citizens and places they belong to and directly act. The strong relationship with place is a commonality of the presented initiatives demonstrating the human scale dimension of resilience practices where interventions are tangible and can be experienced by citizens in their everyday life. The attention to the human dimension, the strong relationship between people and places, the tangible effects of the implemented actions on citizens' everyday life could be considered intrinsical characteristics of the majority of resilience practices as well as the reasons why the transition initiatives are recognized as fundamental resources and opportunities for sustainable urban policies and measures implementation.

The integration and dialog with urban/territorial policies are controversial and not homogeneous in the resilience practices geographies. A lack of large-scale arrangements could be rooted in and confirmed by the emerging diffuse difficulty in establishing alliances among practices and public institutions. The resilience practices of the RPO project (Chapter 5) denounce a shared lack of trust in institutions as a critical aspect. The Transition Network in Spain (Chapter 11) confirms the criticalities that emerged between transition initiatives and institutional actors, even denouncing a diffuse phenomenon of initiatives mistrust in public institutions.

At the same time, several institutional-led processes addressing climate adaptation, environmental, social, or urban innovation process (for instance, in the cases of Milano or Bologna explored by the RPO) were able to activate extensive and inclusive processes engaging and valorizing the practices already existing and active and supporting the activation of practices as crucial factors enabling the achievement of the long-term visions for adaptation or urban innovation and related policies implementation. The resilience practices often refer to European or international sustainability goals as a general framework for local action. However, the contextualization in the territorial scales and an aware medium-scale visioning/framework in which the practice acts/contributes are not consolidated and diffuse.

The openness and transparency of processes emerge newly, as discussed in Chapter 16, as a crucial factor for enabling alliances among resilience practices and public institutions and in boosting the development of large-scale arrangements.

17.2 [Eco]systems of resilience practices: fostering contribution toward sustainability

In recent decades advancement and an increased maturity of resilience practices are noticeable. Considering the multifaceted geographies of [eco] systems of practices from both the disciplinary debate and the reflections presented in the book, a more diffuse recognition of the role that community-led processes play in complex transition processes emerges. The increased recognition is also visible in the active involvement of practices in international decision-making processes as well as the ever greater diffusion of platforms aimed at offering spaces for practices and experimentations networking. A very recent example is the European New

Bauhaus initiative (https://europa.eu/new-european-bauhaus/index_en) that emphasizes the role of citizens activation and experimentation as a core ingredient for European urban models innovation.

At the same time, experiences such as the Transition Network not only reached a worldwide diffusion but began to initiate processes of critical retracement and reorientation to improve the resilience, inclusion, and stabilization of transition initiatives as demonstrated in Chapter 11 on the Spanish case. The Bengaluru case (Chapter 12) shows that the community contribution to medium- to large-scale vision for landscape and ecological values is based on mixed partnerships and arrangements.

The increased recognition of the role of the commons in the transitional process for sustainability positively also impacts the incisiveness of single practices and encourages alliances between different practices and dimensions toward more mature and aware networks and resilience practices [eco]systems.

While underlining advancement, some criticalities emerge (as mentioned in Chapter 16) that require efforts to strengthen the contribution and stabilization of resilience practices.

The awareness of citizens and the engagement of communities are a fundamental principle of the Transition Town Network, reached through training activities, films, and community events that can activate debate and stimulate single citizens and community interest. The innovative aspect is that awareness is generally increased not through a mere transfer of information but by promoting knowledge coproduction, capable of increasing active citizen awareness and supporting the sharing of skills and competencies (Amin & Cohendet, 2004; Berkes et al., 2000; Chaskin et al., 2017; Maesschalck, 2017).

An inspiring perspective is the biunivocal process of knowledge coproduction promoted by resilience practices: sharing knowledge implies recognizing the value of different kinds of knowledge (expert, organizational or institutional, social, and cultural knowledge) enabling mutual exchanges of competencies from different "sources" and resources.

In Transition initiatives, particularly, the citizens' role is also strengthened thanks to communication tools based on the storytelling approach to emphasize that each citizen can "make the difference" and give them the feeling to be part of the change and key players of the transition process.

17.2.1 Perspectives for [Eco]systems of resilience practices

A shared aspect among the experiences presented in the book is the need to activate multilevel and multisector synergies between resilience practices and institutional-led policies, programs, and processes fostering alliances among the resilience practices themselves and among the practices and the more traditional and consolidated institutional and market players.

As much the single practice acts by being aware in being a component of a [eco]system of practices, as much the urban/territorial system benefits can be amplified, leading to vital advancements toward cultural transitions and implementing effective shifts of the regimes of complex urban and territorial systems toward more sustainable and adaptive models. As discussed and anticipated in Chapter 16, the networking dimension constitutes a promising solution, optimal if the network embraces the ecosystem principles and characteristics. This metaphor allows to dissolve some critical issues: the closed or the thematic nature of some networking, configuring the ecosystem as a third and not thematic space where each practice is a functional component, an activating node of synergies and a resource of knowledge and skills.

The two networking cases presented, the RPO project and Transition Network in Spain, although with different roots and characters, show remarkable shared aspects that can be inspirations to outline some perspectives to strengthen the [eco]system networks of practices.

The [eco]systems of practices should preferably be based on nonthematic or sectoral/specialized networking. The network characterization based on a particular theme leads to more specialization in delineating intervention and action solutions and often to a higher focus on technical/solution aspects than on ones concerning inclusion, integration, and rootedness in the local context. As underlined earlier, thematic networks are fundamental in elaborating and experimenting with specific solutions and in considering and promoting sectoral awareness to a broader population stratum.

Not being thematic allows concentrating the efforts of practices capacitation on process strengthening, implementing principles, enhancing the complex approach. On the other hand, as the Transition Towns demonstrate, in the context of networking, structured and informal thematic exchange groups arose demonstrating the dynamic nature of the network. Likewise, to strengthen ecosystem principles and long-term vision, the networking must be as open and inclusive as possible, engaging the local initiatives regardless of the "relevance", the "state of implementation", and the single practice's focus issue.

The Transition Network is an open network. However, it requires an official adhesion to the inspiring principles of the network itself. The RPO network appears as a neutral third space based on a unique partnership including academic actors, the REsilienceLAB association, and financed by a private bank foundation. The RPO defines a set of conditions that have to characterise the practices for their engagement. Both networks are open and inclusive and, at the same time, established conditions for practices adhesion and engagement.

The [eco]systems of practices are based on enabling networks capable of providing the skills necessary to strengthen the quality of the processes of the individual practices. In particular, networks can play a crucial role in promoting cultural transition processes by orienting transversal principles and integrated approaches (Berkes & Ross, 2013).

The capacitation tools are not mainly oriented to developing toolkits for interventions and technical-operational aspects. Nevertheless, they primarily address the capacity building of crosscutting aspects fostering the integration of ethical principles (assuming, as Chapters 1 and 15 presented, the creative diversity and inclusion of all as guiding principles for action) or providing tools fostering the activation of synergies with actors and policies, program, and project of local and territorial scales.

Networks could develop or play a role fostering the transparency and accountability concerning individual practices, guaranteeing, on the one hand, their democracy and consistency with sustainability principles and objectives and allowing practical evaluation and self-evaluation paths as is partially promoted, for example, in the Transition Network. The Lyon case study (Chapter 10) implemented a promising framework supporting the practices accountability supporting both the evaluation/monitoring of the practices and the transparency of the governance process.

The [eco]systems of resilience practices can contribute to enabling incremental and adaptive approaches. The explored cases demonstrated the capacity to orient activities and roles according to the dynamic conditions (rapid changes) by adjusting the action focuses and promoting activities to face emerging needs and urgencies.

Networks (particularly the RPO case) are a space, even physical, for the confrontation and dialog between practices. They constitute a common framework that provides visibility and recognition to individual practices, enables dialogue and exchange between the different experiences and allows the circulation of innovations and experiments. Enabling and emphasizing the practice narration and the comparison with other practices are promising elements to promote critical reflection/self-evaluation and the improvement of individual practice.

The [eco]systems of practices are here interpreted and prefigured as non-thematic, open, enabling networks between resilience practices, the architecture of which should primarily involve the practices themselves. The governance architecture can be maintained fluid and not predetermined. In these networks the actors of academia and research should play a crucial role not only for the richness of the exchange toward innovation and experimentation but also for the social responsibility of all researchers toward an effective and operational contribution of processes toward sustainability. Indeed, the actors and the partnerships, as demonstrated, will have to be open to allow multisectoral and multiscalar dialogs and connections.

Nowadays, the networks and the resilience practices investigated have demonstrated to actively contribute to strengthening the resilience capacity of urban—territorial systems and experimenting and implementing processes toward more inclusive, aware and sustainable complex systems.

The practices have shown capabilities of providing long-term and complex responses facing climate change challenges and towards sustainability and abilities to reorganize their actions according to overcome contingent emergencies (such as the COVID pandemic).

17.3 Closing consideration on pandemic event and the challenges for sustainability (June 2021)

In 2020 the book "l'assemblea degli animali. Una favola selvaggia" was published, which could be translated as "Animal assembly. A wild tale" (Filelfo & Mannelli, 2020). Through the narrative tale style, the book presents a profound, rich, and complex path through the philosophical positions from classical myths to modern approaches that have reasoned on the relationship between humans and the natural world. Along with multiple levels of reading, the book is intense and stimulating in investigating the reasons and philosophical roots of nowadays ecological emergencies and human responsibilities. This book focuses on resilience practices and uses the "ecological" and "botanical" metaphors to reframe complex dynamics and concepts, emphasizing the need to shift all cultural models toward a holistic framework, including the ecological and nature as core components for future visioning.

During the pandemic the reflections deriving from everyday life experiences have solicited questions and activated connections among consolidated research paths onresilience. The Foreword has introduced and outlined points of reflection posed by the pandemic period; this chapter intended to retrace the outcomes of the contributions of cases envisioning ideas for further research paths approached with a renewed and enriched cultural and knowledge awareness.

The pandemic unquestionably demonstrated that health is the result of a complex and multiple synergies of factors involving the quality and healthiness of the environment, the social and economic organization, the individual and collective behaviours. Confirming that health is the result of the interaction between the environment and society, it becomes essential and urgent to include it in policies, strategies, and actions for sustainability and resilience of the territories.

The pandemic has reminded us that we are in a socio-ecological system and, therefore, responses must be systemic. They can no longer be limited to addressing a single problem at a time, forgetting the relationships and the disciplinary cross-fertilization (Nussbaum, 2016). Furthermore, future perspectives must be rooted in synergies and alliances that allow socio-ecological systems to face emerging challenges, rethinking the spatial organization, restoring ecological functions, decreasing the human pressures on the environment, and, simultaneously, providing alternative models for social and economic development. Finally, the panorama discussed shows the crucial relevance to strengthening interdisciplinarity and fostering indispensable alliances between actors of all expertise/sectorial competencies to envisage, implement, and manage projects and transformative processes that effectively pursue sustainability and resilience. Researchers and academia need to commit themself to imagine and act place-based, territorial, and complex visions. Resilience practices show us how local communities can activate transformative processes regenerating territories and cities, contributing to sustainability and resilience capacities improvement.

Acknowledgments and Attribution

The author is grateful to all the authors of the book and all the members of the REsilienceLAB association for their contributions and commitment, especially in providing inspirational ideas and sharing their considerations during the finalization phase of this enriching exchange path that occurred during this challenging period of pandemic crisis.

References

Acuto, M., & Leffel, B. (2021). Understanding the global ecosystem of city networks. *Urban Studies, 58*(9), 1758–1774. Available from https://doi.org/10.1177/0042098020929261.

Amin, A., & Cohendet, P. (2004). *Architectures of Knowledge: Firms, Capabilities, and Communities*. Oxford University Press.

Batliwala, S. (2002). Grassroots movements as transnational actors: Implications for global civil society. *Voluntas: International Journal of Voluntary and Nonprofit Organizations, 13* (4), 393–409. Available from https://doi.org/10.1023/A:1022014127214.

Berkes, F., & Ross, H. (2013). Community resilience: Toward an integrated approach. *Society and Natural Resources, 26*(1), 5–20. Available from https://doi.org/10.1080/08941920.2012.736605.

Berkes, F., Colding, J., & Folke, C. (2000). Rediscovery of traditional ecological knowledge as adaptive management. *Ecological Applications, 10*(5), 1251–1262. Available from https://doi.org/10.1890/1051-0761(2000)010[1251:ROTEKA]2.0.CO;2.

Bixler, R. P., Lieberknecht, K., Atshan, S., Zutz, C. P., Richter, S. M., & Belaire, J. A. (2020). Reframing urban governance for resilience implementation: The role of network closure and other insights from a network approach. *Cities, 103*, 102726. Available from https://doi.org/10.1016/j.cities.2020.102726.

Chaskin, R. J., Brown, P., Venkatesh, S., & Vidal, A. (2017). *Building Community Capacity* (1st ed.). Routledge. Available from https://doi.org/10.4324/9781315081892.

Colucci, A. (2018). *The transition approach in urban innovations: local responses to climate change. Smart, Resilient and Transition Cities: Emerging Approaches and Tools for A Climate-Sensitive Urban Development*. Elsevier.

Feola, G., & Butt, A. (2017). The diffusion of grassroots innovations for sustainability in Italy and Great Britain: An exploratory spatial data analysis: The diffusion of grassroots innovations for sustainability in Italy and Great Britain. *The Geographical Journal, 183* (1), 16–33. Available from https://doi.org/10.1111/geoj.12153.

Filelfo, d., & Mannelli, R. (2020). *L'assemblea degli animali: Una favola selvaggia*.

Hughes, S., Giest, S., & Tozer, L. (2020). Accountability and data-driven urban climate governance. *Nature Climate Change, 10*(12), 1085–1090. Available from https://doi.org/10.1038/s41558-020-00953-z.

Kemp, R., Loorbach, D., & Rotmans, J. (2007). Transition management as a model for managing processes of co-evolution towards sustainable development. *International Journal of Sustainable Development and World Ecology, 14*(1), 78–91. Available from https://doi.org/10.1080/13504500709469709.

Loorbach, D., Frantzeskaki, N., & Avelino, F. (2017). Sustainability transitions research: Transforming science and practice for societal change. *Annual Review of Environment and Resources, 42*(1), 599–626. Available from https://doi.org/10.1146/annurev-environ-102014-021340.

Maesschalck, M. (2017). *Reflexive Governance for Research and Innovative Knowledge. ISTE*. Wiley.

Nussbaum, M. C. (2016). Economics still needs philosophy. *Review of Social Economy, 74* (3), 229–247. Available from https://doi.org/10.1080/00346764.2015.1044843.

Ostrom, E. (1990). *Governing the Commons: The Evolution of Institutions for Collective Action* (1st ed.). Cambridge University Press. Available from https://doi.org/10.1017/CBO9780511807763.

Ostrom, E. (2005). *Understanding Institutional Diversity*. Princeton University Press.

Ravi, R. V. (2019). *Partnership Among Grassroots Organizations*. MJP Publisher.

Roelich, K., & Giesekam, J. (2019). Decision making under uncertainty in climate change mitigation: Introducing multiple actor motivations, agency and influence. *Climate Policy, 19*(2), 175–188. Available from https://doi.org/10.1080/14693062.2018.1479238.

United Nations General Assembly. (2015). *Transforming our world: The 2030 agenda for sustainable development (A/RES/70/1)*. https://sdgs.un.org/2030agenda.

Weber, E. P. (2003). *Bringing society back. Grassroots Ecosystem Management, Accountability, and Sustainable Communities*. MIT Press.

Index

Note: Page numbers followed by "*f*" and "*t*" refer to figures and tables, respectively.

S

Printed in the United States
by Baker & Taylor Publisher Services